Open Distance Learning

Open Distance Learning

Fundamentals, Developments, and Modelling

Oleg Zaikin

WARSAW SCHOOL
OF COMPUTER SCIENCE

Academic Series, WSCS IT and Teleinformatics

JENNY STANFORD
PUBLISHING

Published by

Jenny Stanford Publishing Pte. Ltd.
101 Thomson Road
#06-01, United Square
Singapore 307591

Email: editorial@jennystanford.com
Web: www.jennystanford.com

British Library Cataloguing-in-Publication Data
A catalogue record for this book is available from the British Library.

Open Distance Learning: Fundamentals, Developments, and Modelling

ISBN 978-981-4877-55-8 (Hardcover)
ISBN 978-1-003-13261-5 (eBook)

Contents

Preface

ODL stands for open and distance learning, which is a completely new concept of teaching organization in European Union universities. The implementation of each concept regarding a new way of organizing the functioning of the ODL system (ODLS) needs a precise analysis of its structure as a management object. The complexity and scale of the ODLS determines the development of an appropriate information management system, which combines the features of the traditional understanding of the concept of distance learning and its new approach to open learning. The book is concerned with important and topical problem of ODL, closely connected with the use of modern information technologies in education. The main subject of the book comprises the development of an information management system for the support of the distance learning process in colleges and universities in the context of steadily expanding information technologies. Based on exiting standards (SCORM, etc.) as well as information systems and platforms (e.g. Moodle), the author proposes a coherent and original concept of modelling of three ODL sub-processes: (1) Ontological modelling—how to represent the structure of didactic materials, (2) motivation and collaboration modelling—how to motivate project process participants and (3) simulation modelling—how to realise the learning/project process. In the book, an educational system is represented that can be used as a model of open distance learning process aimed at active behaviour of students and teachers in not only thc dcvelopment of knowledge but also the acquisition of competencies.

Oleg Zaikin
Warszawa
2022

Foreword

The development of technology and civilisation progress has lead to the necessity of general (e.g. spread on the whole population) and permanent learning. In an information society, knowledge will be the top value, and every member of such society must continuously increase possessed knowledge. Lifelong learning (LLL) is no longer a theoretical postulate; in fact, it is an actual imperative, very important for the economical and social development of contemporary countries and societies as well as their civilisation and cultural promotion.

LLL is not achievable on the basis of a traditional school model. Organisations such as primary and secondary schools, colleges and universities are, of course, still necessary but are definitely not enough. For effective spreading of information, for sowing the seed of knowledge, for growing up wisdom, entirely new methods, institutions and tools are necessary. Fortunately, the development of computer and information technologies (CIT) gives now absolutely fantastic possibilities related to so-called distance learning, Internet-based learning, computer-aided learning and many other methods, technologies and tools collectively termed e-learning.

Distance learning is now quite a widespread technology used in many schools for better, cheaper and more intensive teaching and learning procedures. But what we need, in fact, now is an open distance learning system (ODLS), because only open resources can be effectively used for achieving civilisation-oriented goals in the whole society. In this book, ODLSs are described, discussed, proposed and evaluated. We concentrate on the most important and most modern type of ODLSs. For example, we talk about artificial intelligence (AI), which is a very important and very useful element of every ODLS, because the learning process is much more effective if it is planned, organised, performed and assessed in an intelligent way. This AI factor is especially needed in ODLSs because such systems must be able to effectively cooperate with many different learning people— exactly as it is planned in LLL politics. Therefore thinking about the future distance learning system, we must develop ODLSs, improving their concept, perfecting their models, developing their algorithms and spreading their application.

This book contains all the aforementioned elements. It describes basic concept of ODLSs, taking into account the general evolution of learning and teaching systems, as well as considering actual possibilities of CIT. In fact, its approach is based on AI and CIT and founded on system analysis principles. The book discusses various models of ODLSs, derived from different theoretical frameworks: system science, cybernetics, computer technology, content-based, educational-tuned, psychological and pedagogical. A very important element of the book's proposition is related to competence-based system designing, which is treated as being key for competence model building, intelligent teaching and learning system designing and system algorithms development.

Concepts, models and algorithms, as well as detailed methods for achieving particular goals formulated for ODLSs discussed in the book, are presented both theoretically and in the context of practical applications. All applications of ODLSs discussed in the book are based on scientific results collected by the author during over 10 years of practical research performed in the ODL area.

All elements described in the book—concept, models, algorithms, implemented systems and practical experience—are devoted to one goal: to make ODLSs a fundamental technology for the development of a future ubiquitous open lifelong learning system, which should be the answer for the most challenging questions of the 21st-century civilisation.

ODL is an abbreviation of open and distance learning, which is a completely new concept of teaching in European Union universities. Its basic idea was presented in the Bologna Declaration. The implementation of each concept regarding a new way of organising the functioning of the social system needs precise analysis of the structure of the future system as a management object. The complexity and scale of the future learning system determines the development of an appropriate information management system.

Further in the reflections on ODL, the book will use the term 'open distance learning system', meaning an information management system that combines the features of the traditional understanding of the concept of distance learning and its new approach to open learning.

The ODLS is a consequence of the emergence, development and already achieved successes of the Bologna Process. The Bologna Declaration was created in 1999 to integrate the ideas and actions of 29 European countries for the development of the European Higher Education Area, which was to be created by 2010.

The Bologna Declaration does not close the possibility of joining other countries that want to develop together within the framework of the ideas it proclaims. According to the Bologna Declaration, it is necessary to increase the level of standardisation of administrative procedures, among others, in the process of developing programmes and subjects of three-level education and the credit system for student achievement assessment (European Credit Transfer and Accumulation System [ECTS]), which will support the mobility of academic and teaching staff of the university and, most importantly, student mobility.

The extension of the scope of the learning system to the community of European countries necessitates an international-quality standard for each stage of the design and use of distance learning systems and the formulation of a quality assessment for the results of specialist training. One of the main areas of research in the field of ODL is the issue of quality (McGorry, 2004). This problem is discussed as part of the e-Quality project (Quality implementation in Open and Distance Learning in a Multicultural European Environment [2003–2006]), which is financed by the European Union programme Socrates/Minerva. The following institutions participate in the project: the European University Pole of Montpellier and Languedoc-Roussillon (France, coordinator), University Montpellier 2 (France), the Open University of Catalonia (Spain), the University of Tampere (Finland), the Technical University of Szczecin (Poland), the University of Applied Sciences Valais (Switzerland) and Lausanne University (Switzerland). The main goal of the project is to develop standards for creating didactic materials and managing the distance learning environment, while maintaining quality criteria. There is a serious difference in the organisation of the functioning of distance learning systems in the traditional sense and distance learning resulting from the principles of the Bologna Declaration.

The ODLS is the idea of creating such a learning system that will enable, through the information and communications technology (ICT) network, study at the universities of the European Union for every student regardless of his/her current place of residence according to his/her own personalised learning path. The ODLS as a management object is a complex system. Its components, apart from being of a different nature, are often characterised by opposing goals and different laws of behaviour and are mutually dependent on each other. The operation of the ODLS includes individuals, groups of specialists, organisations, computer technologies, the region's and the organisation's infrastructure, information resources and knowledge resources. Implementation of the idea of the ODLS largely depends on the extent to which it is possible to develop an appropriate management system in the form of an information system that will be based on the integrated model of the facility's functioning as a coherent whole.

To sum up, we can assume that the ODLS can be treated as an information system that is designed to manage the process of ODL that meets the conditions of the Bologna Declaration. Success in developing the concept of this information system will allow us to develop a methodology for implementing the idea of the Bologna Declaration in any educational organisation and at the same time will serve as the basis for determining the quality of the organisation of the educational process.

In this book, the subject of research is ODL and is represented as the concept of the operating of an information system designed to support the management process of an educational organisation conducting teaching in accordance with the principles of the Bologna Declaration. The main research method is based on the theory of multi-level hierarchical systems (Mesarovic et al., 1970) that allows you to perform system analysis of the complex system as a whole.

The objective of the considerations in the book is to develop the concept of the ODLS on the basis of the results of system analysis and to show how traditional problems in the organisation of ODL are evolving as part of the proposed concept, including:

1) Modelling of transferred knowledge, acquired competencies and learning materials in ODL conditions

2) The use of an ontological approach as an instrument of independent mastery of procedural knowledge based on appropriate theoretical knowledge

3) Simulation of the learning process as intangible production network

Each of these problems is considered on the basis of the theoretical and practical principles relevant to it. Objective reasons arising from the principles of the Bologna Declaration require a high level of openness in the learning system. This leads to the need to consider the aforementioned problems at the level of a general, slightly formalised description of the system's functioning. In this context, the process of continuous adaptation of the ODLS to the current requirements of the labour market, technology and scientific and research development was specified and described at the meta-knowledge level.

The first problem is analysed on the ontological modelling canvas. Ontology as a method of knowledge representation through specification of concepts and relationships between them gives the possibility to conceptualise a given knowledge domain.

The second problem is studied, on the one hand, in terms of cognitive science and, on the other hand, in terms of motivation modelling. The principles of cooperation of specialists during the development of the project profiles have been described on the basis of game theory.

The third problem concerns the specifics of using the educational organisation's network resources in performing all the processes that make up the functioning of the ODLS. The problem was analysed on the basis of simulation modelling and queuing systems theory.

The most important task of the conceptualisation stage is to formulate a philosophical base that allows you to find the method of analysis that best suits the nature of the object and subject of research. The stage of conceptualising the information system for educational object management, which, in turn, is a complex system, involves an additional complicated task. The layout of the system's components should be determined, and a skeleton scheme of their interaction within the uniform functional scheme should be developed.

Compared to existing information systems intended for management and support of learning processes, the ODLS must be considered as a completely different class system, for which the idea of an open system is an important indicator. The implementation of the ODLS idea will have a major impact on the social situation of society. The importance of the conceptualisation stage increases considerably in comparison with traditional information systems. The lack of a concept will not allow the integration of separate universities into the common European educational system and the creation of a common market for educational services.

In the first chapter, devoted to the presentation of the research object, the scope and operating principles of the ODLS are defined. The basic issues of openness in distance learning systems are presented: openness of the ODLS in the social and IT aspect, openness of the student's life cycle, the concept of the open corporate distance learning network and the principles of developing an open knowledge repository. The difference between the functioning of an educational organisation in the conditions of traditionally understood learning and teaching and in ODL conditions is analysed. Specific didactic materials and the change of the teacher's role in an ODLS are defined.

Chapter 2 presents the approach to competencies in the context of the problem of the distance learning process. The place of the problem of determining competence in ODL is analysed. Competence in ODL conditions is becoming one of the basic instruments for assessing the quality of the learning process. Therefore, it is important to develop the theoretical basis for modelling competencies and adapt this model to practical needs. An approach combining methods of game theory and fuzzy sets would allow determining the degree of competence possessed and acquired by all participants in the teaching process.

In the third chapter, the new educational model developed on acquiring the project team competence is represented. The issue of selecting partners to work on the project profile is considered. The quality of the project team is determined by the degree of coverage of the domain of the project profile by the competencies of its participants.

In the fourth chapter, the problem of how to represent the structure of didactic materials is analysed on the ontological modelling canvas. Ontology as a method of knowledge representation through specification of concepts and relationships between them gives the possibility to conceptualise a given knowledge domain. It is possible to specify concepts describing theoretical, procedural and project knowledge in order to create a hierarchy of those concepts and to show their sequence.

The fifth chapter contains the concept of developing a motivation model aimed at supporting activity and cooperation of both the students and the teacher. The structure of the motivation model and formal assumptions are presented. The proposed model constitutes the theoretical formalisation of the new situation, when a teacher and the students are obligated to elaborate on a didactic material repository in accordance with the competence requirements. The mathematical method, based on game theory and simulation, is suggested.

In the sixth chapter, the author presents the methods and tools for competence modelling for project knowledge management. On the basis of this and also using a fuzzy competence model, the cost estimation model in the form of a group competencies expansion algorithm is proposed. The model focuses on the cost of project team competence expansion caused by the project development process.

In the seventh chapter, the author presents the methods and tools for incentive modelling for project knowledge management. A kind of educational system is represented that can be used as a model of project learning process aimed at active behaviour of students in not only the project development of knowledge but also acquisition of competencies.

Chapter 8 is dedicated to the issues of simulation modelling in an ODLS, which is presented as a production system. It has been shown that the integrated model of a production system in an educational organisation contains three basic sub-models: the enterprise infrastructure model, the production process model and the production network model. Then the task of optimising the structure and parameters of the production network is formulated. The chapter has developed a set of model simulation models describing typical situations of the teacher's work with students during the acquisition of competencies. In addition, an appropriate simulation model is presented to support the student's decision-making analysis during education.

Prof. Emma Kusztina
Warsaw School of Computer Science, Poland

Acknowledgements

All materials of the book, except well-known fundamental principles stated in Chapters 4–8, are original and have been published in the literature for the first time. A considerable part of the material of Chapters 2–4 is borrowed from the book *Intelligent Open Learning Systems: Concepts, Models and Algorithms* by Różewski, P., Kusztina, E., Tadeusiewicz, R., and Zaikin, O. (2011), Springer-Verlag Berlin Heidelberg, Intelligent Systems Reference Library, pp. 1–258.

The author is grateful to E. Kusztina, whose book *The Concept of an Open Distance Learning Information System. Structure and Models of Functioning*, West Pomeranian University of Technology (2006), was used in Chapters 1 and 8. The author would like to thank the president and rector of the Warsaw School of Computer Science, who made a considerable contribution to the publishing of the book.

This research was supported by e-Quality Socrates/Minerva project no. 110231-CP-1-2003-1-MINERVA-M ("e-Quality", 2004).

Chapter 1

Open Distance Learning

1.1 The Concept of Open and Distance Learning

Open and distance learning (ODL) is a new stage of information system evolution in the distance learning domain, the idea of the learning/teaching process organisation in higher education institutions, joining the new teaching technology to expand everyone's possibilities to learn in every life situation (Fig. 1.1).

The main idea of ODL is connected to fulfilling the widely understood mission of the information society. It covers all activities directly or indirectly connected to the process of ODL of different people, students and pupils and provides appropriate infrastructure and legislation. ODL systems are interpreted as the technological basis for the more general concept of distance learning. The concept of ODL will be discussed on the basis of (Kushtina, 2006).

In the report developed by (Patru and Khvilon, 2002) each aspect of ODL was thoroughly analysed. The distance aspect defines the educational situation in which the student is far away in space from the didactic materials and the different participants of the learning process. Communication with the system and other participants occurs only on the basis of the prepared computer environment. Such

Open Distance Learning: Fundamentals, Developments, and Modelling
Oleg Zaikin
Copyright © 2023 Jenny Stanford Publishing Pte. Ltd.
ISBN 978-981-4877-55-8 (Hardcover), 978-1-003-13261-5 (eBook)
www.jennystanford.com

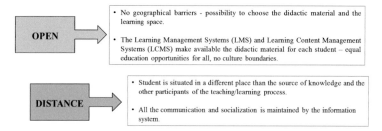

Figure 1.1 Background: open and distance learning (ODL) – new challenges and expectations.

an approach is classified as learning-teaching in an asynchronous mode and is the opposite of synchronous learning based on online lectures. In ODL, on the basis of hardware and software solutions, an individualised, virtual learning space is built. Basing on the methods of artificial intelligence, educational sequences are created and are, on the basis of certain pedagogical methodologies (in the best-case constructivism), passed to the student.

The openness aspect of the ODL process is visible in the strategy and policy that underlies the approach. Each user is to have the possibility to freely choose the material he/she will learn and the place of study. An important aspect of openness is the message behind the entire idea of distance learning – distance learning systems make the same (usually high quality) didactic material available for each participant of the learning-teaching process. Such an approach allows realising the idea of equal educational chances for everyone, which is one of the main factors of developing ODL in Europe.

An important aspect that influences an ODL-based education system, is the change of the paradigm of education, which evolves from the traditional form to the form of distance learning based on the Internet. Introducing this new system, represented by the idea of ODL (Kushtina and Różewski, 2004), will ultimately separate the process of gaining education from the place where the providers of the learning content are situated (meaning the university base) and from the place where the student is present at a given moment (Juszczyk, 2002). Except for special situations, for example connected to the process of certification, this stands for the possibility of breaking up with the three main uniformity rules of traditional education: uniformity of place, action and time. As the challenges connected to the mobility of European citizens require education to be organised in such a way that the learners and teachers could meet in the information space (so-called cyberspace), the uniformity of place does not have to occur (Kushtina *et al.*, 2001). Professors can provide the ODL system with a learning content in place X, and their students can use this and other contents in places Y_1, Y_2,..., Y_n. Moreover, no two places need to be identical, and a functioning connection of the professor with each of the students can still be fully guaranteed. Also, without much hassle, the conditions of a constant, close communication between the students can be ensured, even if each of these students is in a different place or travelling at the moment.

Similarly, it is not necessary to maintain the uniformity of time. In the traditional learning process, the lecturer and all the students listening to the lecture have to perform actions connected to their roles exactly at the same time. Students who come late to a lecture irrevocably lose some part of it. It does not have to be this way anymore. The lecturer can supply the ODL system at the moment that is most convenient for him/her; the student can derive knowledge from the ODL system when he/she is best prepared physically, mentally and organisationally; and all interactions (greatly desired in the learning process) can occur completely asynchronously (Greenberg, 2002).

1.1.1 Aspects of Open and Distance Learning in the Scale of European Union

To fulfil the concept of ODL, the following elements should be discussed in the context of the learning systems:

- *Social aspect*, the widening of which results mainly from two causes: Firstly, the area of the system's functioning is no longer limited to a certain group of people but to an entire community, in the frames of the weakening meaning of geographical borders. Secondly, quick deactualisation of domain knowledge and changes in the working conditions result in a social need for lifelong learning.

- *Cultural aspect*, which covers managing the process of diffusion of cultures and national traditions with the need of creating European values, while maintaining the status quo for each ethnic and social group.
- *Technological aspect*, which covers the issues of assessing the scientific, technical and organisational possibilities of supporting the functioning of ODL as a system for managing the planning, preparing and providing of a personalised educational service.
- Broadening of the *economical aspect* results from the need to increase the required level of complexity and universality of hardware and software resources. This activity increases the cost of ODL, which is contradictory to the idea of fulfilling social requirements that assume lowering costs of education with a simultaneous increase in its quality. Finding a compromise becomes the main economical problem.
- Arising of the *political aspect* is caused by the fact that ODL, which crosses national borders with its scope, requires direct involvement and participation of not only European Union (EU) bodies but also of the governments of all involved countries.

The need to include humanistic values, personalisation of educational services and the scale of popularisation, qualify ODL as a social system which requires the formulation and appropriate interpretation of certain philosophical values, that is, emergency, synergy, holism and isomorphism.

In general, emergency is the phenomenon of creating a new, previously unmanifested value as a result of the system functioning in a new environment. The new value appears as a result of cooperation between the elements of the system, the characteristics and parameters of which are already defined. This new value becomes a characteristic of the system as a whole. In regard to ODL, what should become such a new value is universal for the entire education system form of knowledge representation as a shared resource, and the semantic model of describing the system of managing the learning-teaching process, reflecting not only its flow but also the content and quality of knowledge obtained by the student at each stage of education.

Synergy is the effect of strengthening each part and the entire system as a result of their cooperation. In the context of open learning systems, apart from traditional planning, organisation and administration of the education process, the issues of volume, depth and structure of knowledge become the object of management. Synergy in this context is expressed through interaction and cooperation, with a certain level of applied knowledge. The summary competence obtained by all participants during the execution of the education process will express the increase of the intellectual capital of the educational organisation working according to the ODL principles.

By *holism* we denote the evolutionary process of creating new entities, considering that the value of the whole has preponderance over the sum of its parts. In that sense, the process of gathering and assimilation of the shared knowledge resource in the frames of big educational organisations of the world obtains the effect of holism. Knowledge resource can be presented as a multi-level structure, where the top levels correspond with the higher, and lower levels with the lower, level of abstraction and generality.

We are dealing with *isomorphism* when the form is system-creating, forming the structure, with greatly different content filling it. Isomorphism is the measure for the appropriateness of structures and functions of objects. In reality, the phenomenon of complete isomorphism does not exist. However, in cognitive processes, where we use abstractions, isomorphism allows explaining the sense of a concept through structuring of its definition. In regard to ODL, isomorphism is the effect of maintaining the structure and functionality of the system at different scales of implementation. Scalability parameters in this case are often quite different: quantitative parameters of the educational organisation, domain knowledge and cognitive styles of teaching.

The mentioned philosophical rules are the essence of maintaining an open learning system (as a uniform object), which defines the need to include them in the model of managing every educational organisation involved in the idea of ODL.

1.1.2 Analysis of the Openness Issue in Distance Learning Systems

In learning systems, the concept of openness can be considered regarding different aspects. Changes that occur in the educational policy of the new EU members require a more specific outline, in a way that allows for defining a uniform approach to the concept of openness.

An example of an open system close to the problems of education is the natural language. Adding new words as a result of different causes (e.g. new technologies, expanding one's own environment of life) does not disturb the main function of the language, which is ensuring communication in situations previously undefined. The main requirement for maintaining this functionality on an unspecified interval of time is the possibility to use the language to formulate new thoughts on the basis of the already stated ones. Following the ontology of thought written in a verbal and symbolic way may serve as a basis for modelling a common resource of knowledge for different groups of people.

1.1.2.1 Openness of the education system in a social aspect

Openness at the social level means the possibility to flexibly adapt the education system to changing social requirements regarding educational services and providing each citizen with the right to increase their competences through the use of the education system in any mode.

The main area of functioning of the education system is society as a whole and each receiver of the educational services individually. The leading idea of open learning is connected to fulfilling the widely understood educational mission of society. The mission covers all activities directly or indirectly linked to the process of distance learning of different groups (including handicapped) of students and pupils, providing appropriate infrastructure and legislation.

The higher the cultural level of a country, the more demanding the bylaw of education (scope of the programme, time and mode of learning, certification conditions). One of the goals of standardisation is to increase the average intelligence resource level of the society of countries included in the Bologna Process. Introducing standards does not aim at abolition of cultural and social specificity of individual countries but guarantees a successful exchange of students, study programmes, didactic material or conducting internships abroad.

In the social context we have to talk about mutual openness of the human and the society for evolutionary development of humanistic and social values. For a person this means that he/she should be prepared for continuous education throughout his/her entire life. For the environment this means continuous adjustment of the infrastructure, legislations and social security.

1.1.2.2 Openness of the learning system in the IT aspect

Distance learning systems are the technological basis for development of the open learning concept. In the scope of computer systems, an open system is a piece of software with flexible structure, where the relations between modules are described in a formal language, and it is possible to add new modules and set (in specified range) the parameters of functioning of each module, according to the implementation conditions. The number and content of the modules to a great extent depends on the goal and criterion of using the system by the distance learning organisation.

Analysis of individual cases of implementation of distance learning systems shows that the distance learning system is understood as a computer (IT) tool, which is in control of several, differently educed processes (i.e. managing student accounts, managing courses, facilitating communication between participants of the education process, reporting, doing statistics, supporting education in the frames of a single topic/course, using simulators for obtaining different vocational abilities). A strict list of these processes, their classification, executing rules for distance learning systems, has not been developed yet.

Analysis of trends in the openness of education systems regarding the computer aspect shows that at the moment solving this problem depends less on hardware and network issues and more on the possibilities of formal representation and manipulation of knowledge presented in the education process.

1.1.2.3 Openness of distance learning corporate networks

Each university included in ODL can be treated as a corporation working as a distributed organisation. Corporate networks are usually used to create a uniform network environment in which the following functions are performed: planning, managing production, administration, logistics and client and staff management.

The market environment and the competition force the university to obtain characteristics of an enterprise and apply its own methods for maintaining the position on the market. Methods of using network resources play an important role in this case. One of the main characteristics of an educational enterprise is the fact that it is oriented on serving a random stream of orders for educational services on the basis of intangible production (Korytkowski and Zaikin, 2004). Each workplace is located in the network, the product/half-product is created as a result of cooperation between all participants of the education process, and it is continuously sent over the network. The change in the role of the teacher, the necessity of his/her cooperation with the new body of specialists (e.g. expert, knowledge engineer) and the constantly changing network environment require continuous uplifting of the qualifications of the staff. This process is conducted more and more often through the methods of distance learning. Therefore, an educational organisation additionally acquires the characteristics of a learning organisation (Zaliwski, 2000).

The aforementioned characteristics of an educational organisation working in ODL conditions show that this organisation uses the features of the corporate network in a much wider sense.

The open corporate network is a computer-based way of realising cooperation, and information and knowledge exchange between members of the organisation and the distributed structure. The main requirements of the distributed organisation towards the corporate network are the following:

- *scalability* – the possibility to change, at any moment, the number of clients and the functional software packages;
- *high efficiency* – the ability to process high amounts of information; guarantees quick performance of complicated applications based on the client–server interface or grid

- *elasticity* – the possibility of automatic or semi-automatic adjustment of the corporate network to often-changing conditions of using the network within the distributed organisation

The corporate approach allows for connecting the requirements of openness of systems in the social sense with the openness in the computer sense. Planning and development of the corporate network requires considering this problem in the context of managing the entire educational enterprise, based on intangible production.

1.1.2.4 Openness of the student life cycle

The student life cycle is a process of cooperation between the student and the university; therefore the issue of openness should be looked at from both points of view. An open student life cycle means that each student has the possibility to choose the learning programme on the basis of certified education offers of European universities, according to their own point of view, but in accordance with the defined rules. The basis for such an approach is the European Credit Transfer System (ECTS).

The main rule that allows for choosing individual programmes of studies, in an ideal situation, should be universal for all universities' taxonomy: specialisation – profile – subject. In the frames of this taxonomy, a relationship of many-to-many should be defined between the specialisation and the subject, which means that one subject can be a part of study programmes for different specialisations, and different specialisations can have common groups of subjects.

The next rule assumes that regardless of what education programme the university offers for teaching a certain specialisation, each subject passed within it has a weighted indicator according to which the student obtains a certain amount of commonly accepted points (ECTS points). Another rule considers defining a minimal sum of points that each student has to obtain during studies following his/her own education path. A given sum of points is the basis for awarding a certificate (diploma) of the specialisation.

The last rule assumes there is a possibility to compose one's own study programme by the student in such a way that he/she will pass individual blocks of the programme at different universities.

The document facilitating recognition of the points obtained by the student will be the diploma supplement.

The student life cycle in the traditional learning conditions is a process of obtaining knowledge according to a specified education programme that clearly defines the content and schedule of this process at one university. Distance learning allows for a freer schedule, performed also at several universities. The student's strategy will then boil down to simply choosing a university that guarantees achieving the appropriate certificate – the student's object of interest. After choosing a university, the student is included in a deterministic student life cycle. From the point of view of the university, the process of servicing all students is a deterministic one, and teaching certain subjects to different groups of students can happen in parallel. The process of carrying out a defined study programme can be described by the university in the form of a Gantt chart.

In the ODL conditions, the student life cycle is a stochastic process. The student, while maintaining the same object of interest, can choose subject (courses) at different universities. The student is able to check at which universities the proposed study programme guarantees obtaining the proper certificate. He/she then independently chooses at which universities to realise his/her own student life cycle. In other words, students move between universities, realising a personalised education path. The structure of the student life cycle is constantly open. After passing a certain block of subjects at a chosen university, the student again faces the situation of choice as to where to continue the education in the future. In this case, on a certain horizon of planning, each university will be dealing with a random number of students at each subject included in the learning programme. In the described conditions, the process of teaching students becomes distributed in the common education space. The education space can be presented in the form of a mass-servicing network, where each node represents a university. Functioning of the so-described process can be presented using a stochastic model of the mass-servicing network (Jackson's model). The education process in each of the nodes of the entire network can be interpreted as pipeline production. Table 1.1 shows a comparison of functioning of distance learning systems and ODL systems from the point of view of servicing students.

Table 1.1 Comparison of functioning of distance learning systems and open and distance learning systems from the point of view of student processing (Kushtina, 2006)

Object of investigation	Distance learning (single institution)	Open and Distance Learning (institutions' network)
Learning curriculum for student of j specialisation (P_j^{ST})	The learning curriculum is compatible with university u curriculum j: $P_j^{ST} = \delta_j^U(P_j)$, δ_j^U – the probability that student selects the curriculum j from the university u	The student's curriculum consist of the courses from different universities: $P_j^{ST} = \delta_{j_1}^{U_1}(p_{j_1})/\delta_{j_2}^{U_2}(p_{j_2})/.../\delta_{j_n}^{U_k}(p_{j_n})$, $\delta_{j_n}^{U_k}(p_{j_n})$ – the probability of student selection of specialisation j on the university u, $p_{j1},...,p_{jn}$ – curses of j specialisation, $U_1, U_2,...U_k$ – universities caring the j specialisation
Infrastructure for learning process	Virtual organisation	Distributed ODL organisation (characterised by the stochastic student's flow)
Student's servicing model	Queuing system($M/M//N/\infty$)	Open queuing network with stochastic transmission (Jackson network)

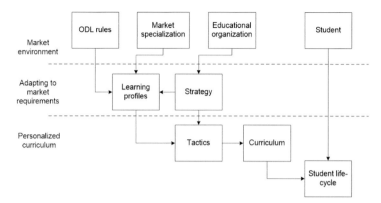

Figure 1.2 Adapting the educational organisation to market requirements (Kushtina, 2006).

Figure 1.2 presents how ODL works, with the market of required specialisations and the potential contingent of students found outside of the educational organisation, creating its surrounding, to which it should comply. Adjusting to the market requirements in the organisation is conducted in two stages: in the beginning through development of appropriate profiles, which the organisation offers on the market of educational services, followed by the development of a study programme according to which the order of a student interested in the offer can be processed. In order to maintain the position on the educational services market, the educational organisation should constantly ensure openness of the student life cycle.

1.1.3 Conclusions

The presented concept of distance learning, compliant with the Bologna Process, undergoes constant evolution caused by the difficulties in adapting the presented idea to the changing conditions of the education market being created. The result of this is the organisation of regular meetings (until now the meetings happened in Bologna in 1999, Prague in 2001, Berlin in 2005, Bergen in 2005, London in 2007 and Leuven in 2009) of education ministers of the countries that participate in the Bologna Process in order to introduce required corrections and specify the strategy for the next period of time.

ODL defines the goal structure, which represents coexistence of the education market and the requirements and social mission behind the education process. In the future some form should be found that will include in its dimension commercial educational solutions together with state education. The vision of European education is additionally supplemented by distance learning, which will become a significant complacement of education conducted in the traditional mode.

1.2 Management Model of Open Distance Learning System

1.2.1 Introduction

The base for functioning of an information system meant for managing an organisation is an integrated model describing the processes occurring inside the organisation and the cooperation between the organisation and its environment. There are many approaches to describe each of the aspects of the organisation's functioning:

- Formal description of the organisation's structures (organisation theory)
- Description of the behaviour motivation of the participants of the processes taking place in organisations (games theory)
- Description of the organisation's activity through modelling the basic functions (systems theory)

Each of the mentioned approaches enables a deep analysis of the organisation's activity; however, none of them considers all of the processes occurring in the organisation and influencing its life cycle at the same time. System thinking makes it possible to distinguish the main processes, to describe their interactions and to find a methodological approach to creating an organisation management system that ensures the best organisation existence conditions.

As an object of research, distance learning consistent with the idea of the ODL (Patru and Khvilon, 2002), carried out by any type of institution, is a system as it possesses a specified goal and a complicated structure that consists of a few subsystems having their own sub-goals and cooperating in the name of a common global goal. In the later part of the book, when talking about ODL we will use the term *open distance learning system (ODLS)*, meaning an information system for distance learning that joins the characteristics of the traditional understanding of the term 'distance learning' and the new understanding of the term 'open learning'.

As a theoretical and methodological base for open system of distance learning (OSDL) analysis, we will use the approach developed by (Mesarovic *et al.*, 1970) in *Theory of Hierarchical, Multilevel Systems*. In that work, the authors introduce the idea of a multilevel hierarchical organisation structure that considers different functioning aspects of the organisation. The hierarchical multilevel analysis is popular in literature, for example, modelling social systems (Miklashevich and Barkaline, 2005) and multi-dimensional complex software systems (Gómez *et al.*, 2001). System analysis, according to this approach, considering the hierarchical nature of organisation management in three dimensions:

(1) Specifying areas of abstract description of the organisation
(2) Distinguishing the layers of decomposition of the problems the organisation is facing
(3) Deciding the order of decisions in the decision-making process when solving problems

In (Kushtina and Różewski, 2004) the hierarchical structure of OSDL was defined in detail. The OSDL is divided into three main subsystems: strategic management system (SMS), learning management system (LMS) and learning content management system (LCMS). Each of the subsystems includes several modules related to different information system functions. For example, the LMS covers: e.g. registration, administration and corporate network issues. The next result of the system analysis should be the functional scheme imitating the placement of the modules according to the educational organisation management cycles.

1.2.2 Open and Distance Learning

The concept of ODL is a completely new idea of organising the learning process in higher education institutions in the EU. Its main idea was presented in the Bologna Declaration (1999). During consecutive meetings of the EU governments in Prague (2001), Berlin (2003) and Bergen (2005), the idea of ODL has been continuously developed. The report created by (Patru and Khvilon, 2002) shows a detailed analysis of every aspect of ODL.

The *distance* aspect describes an education situation where the student is situated far away in space from the didactic materials and other participants of the teaching/learning process. Communication with the system and other users can only take place in a previously prepared interaction environment based on telecommunication technologies. Such an approach is classified as the asynchronous learning mode.

Research in the area of pedagogy, psychology and cognitive science (Anderson, 1995) (Anderson, 2000) has led to a deeper understanding of the nature of knowledge adaptation during the learning process, thus making it possible to introduce the idea of a learning object (LO). In the asynchronous learning mode, the structure of didactic materials is module-based. Every domain can be divided into modules consistent with the distance learning standard SCORM ("Advanced Distributed Learning Initiative", 2001) that can be later used in different courses, without the need for expensive recoding or re-designing operations. In OSDL on the basis of the LO approach, a personalised virtual learning space is being built. Basing on artificial intelligence and cognitive methods, educational sequences are created and then passed to the student.

The *open* aspect of ODL is visible in the current strategy and politics of every EU member country. Each user should have the possibility to freely choose the material he/she is going to learn and the place where the learning process will take place. A more important aspect of being 'open' is the concept also standing behind the whole idea of distance learning – distance learning systems present exactly the same didactic material (usually of high quality) to each and every participant of the learning process. Such approach makes implementing the idea of equal education possibilities for everyone possible, and it plays a major role in developing ODL all over Europe.

The process of building European distance learning systems and structures has already begun many years ago. The analysis made by (Tait, 1996) shows the evolution of distance learning systems that took place in European countries. Consecutive steps undertaken by European governments have gradually reorganised the approach and the point of view at ODL. In the initial period, distance learning in Europe was represented by a set of individually and separately working universities, each of which provided independent services (e.g. Open University, operating in Great Britain since 1969; Fernuniversitat, functioning in Germany since 1974). In the next step, the organisations and institutions teaching over distances began to join each other, creating consortiums focused on specific areas of Europe or groups of interest. Officials and decision-makers of the EU in their consecutively published working and official documents have been gradually building the structure of the mission that distance learning systems and organisations are to fulfil. The main problem the EU raises before a system-based solution to the education issue (based on the distance learning technology) with is equalising the education possibilities of all EU citizens. The education process should promote and develop an understanding of the European cultural heritage. The platform joining all EU citizens in one organism of knowledge civilisation will be built by creating a mutually understanding society.

1.2.3 Hierarchical Structure of ODL

At the stage of conceptualising the system, the proposed detailing of processes is sufficient. It enables characterisation of the type and source of information and defines the principle of information processing to the form of data or knowledge. Further specification of processes should be carried out at the design stage of the information system and requires consideration of the chosen methodology to identify the detected processes.

Thus, the use of the abstraction zone dimension and subsequent specification of the processes that support each layer allowed to distinguish many modules of the information system. Subsequently, it is necessary to arrange these modules in a functional diagram (Fig. 1.3).

The OSDL issue is being studied at the level of information systems in order to create a management model that will enable both further system development and control of its everyday work, which consists of providing educational content to a certain contingent of students. The presented area of operation allows concentrating on the OSDL system's dimensions, which can be joined through the common denominator of information systems. OSDL is a system complicated enough to make building an isolated, 1D model impossible. Therefore, like in every complex system, appropriate subsystems should be distinguished. The result of the OSDL problem analysis is shown in Fig. 1.3.

The global OSDL operation (functioning) criterion can be interpreted in the context of ensuring learning environment conditions which:

[I] Maximise fulfillment of individual student's requirements regarding time and the learning mode.

[II] Minimise differences of the traditional learning environment.

[III] Maximise possibilities to gain a studying (education, learning) results certificate.

A student studying over a distance has certain needs regarding time and the learning mode (aspect I of the global criterion). The most flexible way of solving this issue is developing a dedicated solution based on an information system including a knowledge base which will ensure access to didactic content through the Internet (at any time). However, in the didactic process we should also take into account the social aspect of learning, which assumes interaction with other people, like the teacher (advice, consultation) or other students (discussion groups, joined projects). This extracts developing a system that provides distance students with the best-possible access to information resources of the system but also enables easy contact with other people engaged in the didactic process in the frames of OSDL. This problem is discussed in detail in the section 'Distributed Interactive Learning' in (Khalifa and Lam, 2002).

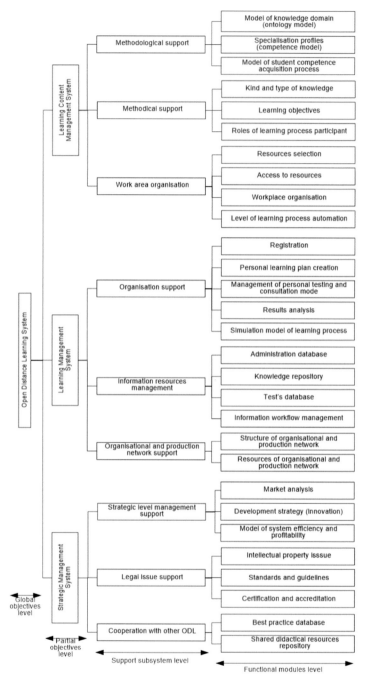

Figure 1.3 Structure of the open distance learning system (own study)

The model of reference of each interaction in the frames of the created virtual space is functionally adequate and equivalent to the model of interaction taking place between people in reality. Therefore, the aim of every OSDL system is to come close to the ideal model of traditional learning (aspect II of the global criterion). The verbal information exchange process is formalised at the level of knowledge manipulation. Knowledge exchange that takes place between traditional learning process participants on the basis of natural language is moved to a limited environment of computer systems, where every form of communication is based on manipulating structures.

The sense of existence of OSDL systems in the context of the existing education system is appointed by the possibility to certify learning results (aspect III of the global criterion). Placing educational solutions in legislative reality forces finding appropriate legislations and legal solutions that allow OSDL units to certify studies on a similar basis as it takes place in the case of traditional institutions.

The present state of knowledge and the characteristic of existing ODL organisations allow determining, in the form of a hierarchy, the set of subsystems that an OSDL system consists of (Fig. 1.3). Individual elements of hierarchy are distinguished and ordered according to technological and scientific rules that characterise them. Each subsystem is described by a partial criterion, tightly connected with the global criterion through common parameters. Distinguished elements are not easy to formulate, because they are characterised by many different parameters of a complex nature. Defining and modelling them requires specialists of different domains, which leads to the level of functional modules.

1.2.3.1 Learning content management systems

The purpose of a learning content management system (LCMS) is to maximise conformity with knowledge exchange environment requirements. Solving such a problem requires analyzing such aspects of learning as culture, organisation, language (information and knowledge exchange) and learning standards (national and international). The LCMS subsystem criterion being considered makes aspect II of the global criterion – minimisation of differences with a traditional learning environment.

The main impulse that leads to developing an LCMS was creating a new approach to building didactic materials meant for asynchronous learning. Distance learning courses are characterised by a strict structure and a closed context. Research in the field of pedagogy, teaching, psychology and cognitive science allowed a better understanding of the nature of knowledge in the learning process, enabling introduction of a module method of didactic materials organisation. As many scientific studies show, e.g. (Maruszewski, 2002), knowledge has a modular form consisting of concepts joined with relations. It is possible to divide every domain into modules according to its structure. On the basis of the module structure of knowledge, we divide didactic material into objects, each of which contains some portion of knowledge. It has been observed that courses belonging to a certain class (e.g. class for higher education algebra courses) have common, identical areas. This means that for every class of courses, there are portions of knowledge invariable in the entire spectrum of courses in that class. Due to that fact, it is possible to design a module once and use it many times in different courses. Knowledge modules, which a given domain is divided into, are called *Learning Object (LO)*. An LO can be interpreted as an idea that allows for multiple usage of one didactic material in different courses, not only of one domain but also of one paradigm. Basic studies of the LO idea can be found in (Downes, 2001) and (Hamel and Ryan-Jones, 2001). Suggestions of methods of modelling knowledge according to an LO idea can be found in (Kushtina and Różewski, 2003), (Lin *et al.*, 2003) and (Wu, 2004). The problem of LOs is also being analyzed by many research and government organisations in the frames of developing SCORM standards ("Advanced Distributed Learning Initiative", 2001).

A distance learning environment does not allow using natural language in the full extent as a tool for exchanging knowledge. Therefore, it is necessary to develop a special *communication language* and dialogue scenarios; a discussion of this issue can be found in (Kushtina and Różewski, 2003). Creating didactic materials based on LOs requires *learning domain modelling* at the level of knowledge. While building a certain LO, we manipulate concepts to create a conceptual model of a given knowledge area. In the best case the created model is adapted to the student's cognitive structures and through assimilation/accommodation operations added to already existing knowledge structures.

One of the key elements of distance learning is orienting the material being developed at a specific *scope and type of knowledge*. Knowledge as an object and aim of learning can be divided into two basic types, fundamental-theoretical and operational. Fundamental knowledge reflects conceptual thinking, the basis for formulating new paradigms, problems, tasks, behaviour rules, etc. Operational knowledge is necessary to realise scenarios, algorithms of performing operations. Situations that people encounter in everyday life are based on simultaneous usage of both types of knowledge in different proportions, depending on the level of task complexity. The *education goal* defines the relation between fundamental and operational knowledge in each case. In the context of educational challenges set by a given education goal, individual user roles are defined. The expectations are different in the case of a flight simulator (a more detailed discussion in (Popov *et al.*, 2003)) and in the case of a basic mathematics course.

For an OSDL system it is characteristic to pay special attention to learning space and individual learning components – *intellectual learning environment organisation*. In a traditional classroom, due to the contact with a 'real' teacher, requirements towards didactic materials and textbooks are not always high. In distance learning didactic materials and the working environment have a very high influence on the quality of learning as they directly create the intellectual didactic transmission process. While designing an *intellectual workplace*, we create a network knowledge-exchange space, choose software while having in mind the specified use of the system and decide the *level of process automation*. An example of a working space in the mathematics domain (called CyberMath) which allows interaction with mathematical models is presented in (Knudsen and Naeve, 2002). Analysis of a session of students working in CyberMath shows that the main factor is achieving a high level of integration with the prepared environment; a lack of *access to materials* or lacks in the environment itself successfully restrict the possibility to achieve the desired feeling by the student.

1.2.3.2 Learning management systems

Learning management systems (LMSs) allow companies (organisations) to plan and track the educational needs of their employees (students), partners and clients. Following (Greenberg, 2002), we can say that LMSs are a strategic solution meant for planning, delivering and managing all educational events within the company (organisation), including virtual classes and classes led by an instructor. LMSs allow *registration and identification* of students through the structure of profiles, which describe individual characteristics of the student and his/her didactic achievements. On the basis of the profile *learning scenario development* becomes possible. Its global aim is to acquire the necessary level of competence. The learning process is supported by *results analysis* tools. On the basis of the analysis of each course, the level of charging teachers, the student's profile and the course's requirements towards the student, *determining individual testing and consultations mode* is possible.

The main LMS parameters are availability and capacity of the learning network with a cost restriction. This connects the issue of the LMS with aspects I and III of the global criterion. Increasing the quality of educational materials availability causes an increase in the comfort of work and enables building a more efficient working space for the student.

An LMS, due to the complexity of the distance learning process, requires access to different types of data and information. The basic base that regulates the work of every organisation is the *administration base*. A *tests database* and a *knowledge repository* are specific for distance learning. The tests database, except for storing the content of tests, also enables adapting a given test to a specific educational situation through ensuring a methodology of adapting tests to a given profile. The method of grading students can be based on a set of developed heuristics or on a specified set of rules. A knowledge repository enables defining a common knowledge model. The knowledge model contains a formal definition of concepts that can be used for modelling knowledge in the given domain and rules that allow creating real statements in the given domain. The knowledge repository can be built according to (Neches *et al.*, 1991) on the basis of collecting unified ontologies in the form of a library. Due to a large number of data and information being sent, it is essential to provide mechanisms of *control and manipulation of the data and information flow* in order to manage restricted resources (bearing in mind priorities of certain tasks).

The national ODL consortium integrates different solutions, which is caused by the specific character of each didactic institution. The big organism of a university each time evolved in a different way, and procedures, documents and basic data are different. It is a big challenge to build a corporate network not only at the level of service but also at the level of information. A *corporate network structure* can be based on the star architecture with the main centre, supplied with all the data from satellite centres, in the middle; further discussion of this issue can be found in (Zaikin, 2002a). Building an ODL consortium requires creating a fast network infrastructure that ensures safe communication at the level of documents and coupling databases and management systems. In Polish reality, the initiative called PIONIER: Polish Optical Internet, described in (Binczewski *et al.*, 2001), is a part of ODL *corporate network resources* and makes available advanced applications, services and technologies adapted to the idea of the Information Society.

1.2.3.3 Strategic management system

Every OSDL organisation functions at an educational market characterised by a high level of competition additionally increased by the opening of Europe for students. Strategic management systems (SMSs) help in defining development strategies in the short-, middle- and long-term context. A strategic management module provides tools and methods for taking decisions about creating a course basket with consideration of potential students (clients), the competitive environment and general social, consumer trends. *Marketing analysis* tools allow the ODL organisation to form OSDL according to new market conditions in order to maintain a high level of competence of the graduates, while also maintaining a high level of competitiveness of the institution. On the basis of marketing analysis and the adapted organisation's mission, the *development strategy* is determined, which in all modern universities (not only of ODL type) adapts the form of a business plan. All considerations about the future of an organisation need a real guarantee of success for the purposes of design and implementation of the project. An approach to creating the model is covered in (Zaikin, 2002a).

The operation of the university, due to its educational mission and social value, should be strongly anchored in the existing legal system. Observance of intellectual property rights in the case when the teaching material is fully digital is a difficult issue. Digital material can be easily copied without loss of quality. Therefore, legal and technological mechanisms are necessary to protect the didactic material. An example solution based on an electronic license and public key structure is presented in (Santos and Ramos, 2004). An equally important aspect of the issue of intellectual property rights is the problem of the use of material, data or information prepared by other authors as part of a course designed by us.

According to EU findings, each country is responsible for the quality of learning individually. The EU provides assistance by providing studies, guidelines and tips. Each country develops its learning standards in accordance with European guidelines, taking into account individual characteristics. Standards do not focus strictly on the learning process, but also on the dean's office (ISO) management process, data security, etc. In the case of distance learning, there is a large group of standards operating at the operational level that regulate the ODLS. The most important standards for distance learning are the Alliance of Remote Instructional Authoring & Distribution Networks for Europe (ARIADNE), the Aviation Industry CBT Committee (AICC), the Institute of Electrical and Electronics Engineers – IMSE LTSC, the IMS Global Learning Consortium Inc . (IMS) and Advanced Distributed Learning (ADL). A detailed discussion of the standards of distance learning with particular emphasis on learning flooded in asynchronous mode can be found in (Friesen and McGreal, 2002).

Certification of distance learning centres may take place at various levels. At the national level, the State Accreditation Commission visits individual educational institutions, assessing the level of education and the level of staff preparation. On the basis of the analysis of the committee's report, accreditation is granted. The procedure is the same for both traditional and ODL universities. There are also industry accreditation institutions, e.g. the Accreditation Committee of Technical Universities (www.kaut.agh.edu.pl) or the Foundation for Promotion and Accreditation of Economic Directions (www.fundacja.edu.pl), whose certificate is prestigious and confirms the high quality of education in the institution.

Organisations teaching over a distance cooperate with each other on different levels. The basis for common operation is a consortium of traditional institutions that would finance together the distance learning initiative. The next step of cooperation is creating a common *best-practice database* and a *shared didactic materials repository*. The best-practice database allows efficient solving of problems that come with implementing new distance learning technologies, which often requires adapting complicated systems and modules, and methodologies to local organisation characteristics. This creates a need for exchanging know-how. The best-practice database is realised in the form of a knowledge repository oriented at practical knowledge. The experience of many ODL institutions shows that the most important element of each ODL organisation is the developed didactic material. Therefore, it is advisable to create cooperation space for ODL organisations, which will enable sharing materials in the frames of a global didactic materials repository. (Galwas, 2003) suggests creating a bank of subjects studied over the Internet in the frames of Polish ODL initiatives.

1.2.4 Functional Scheme of Open System of Distance Learning

Developing an information system dedicated to managing an education organisation is a complex task. In the case of the ODLS, the analysis made by the authors leads to the identification of four embedded management cycles, which differ in their operating time. As can be seen in Fig. 1.4, each cycle includes a process that is being arranged by a certain decision-maker. Within the time limit of each cycle, the system decision-maker compares knowledge areas in order to take decisions by estimating their content and depth (helpful algorithms have been proposed in (Zaikin *et al.*, 2006)).

Figure 1.5 shows the functional scheme of an ODLS, which consists of four inbuilt management cycles. The functional scheme can be described as a process of sequential knowledge processing during (i) syllabus preparation, (ii) providing education services, (iii) developing didactic materials, (iv) the process of acquiring competences based on a specified knowledge model and (v) statistical evaluation of students' progress.

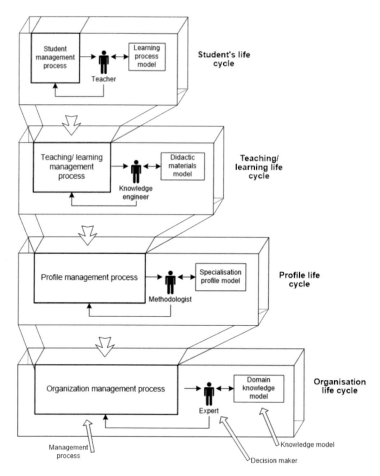

Figure 1.4 Conceptual model of the open distance learning system (ODLS).

1.2.4.1 Knowledge models

In each management cycle, the decision-maker uses an appropriate knowledge model to take the decision. Between the knowledge models and education organisation management subsystems located in the functional scheme (Fig. 1.5), the following relationships occur:

1) *Domain knowledge model (DKM)* relates to the SMS, its structure can be described with the following tuple:

$$DKM = \{Pr, R, Ac, Kd\}, \tag{1.1}$$

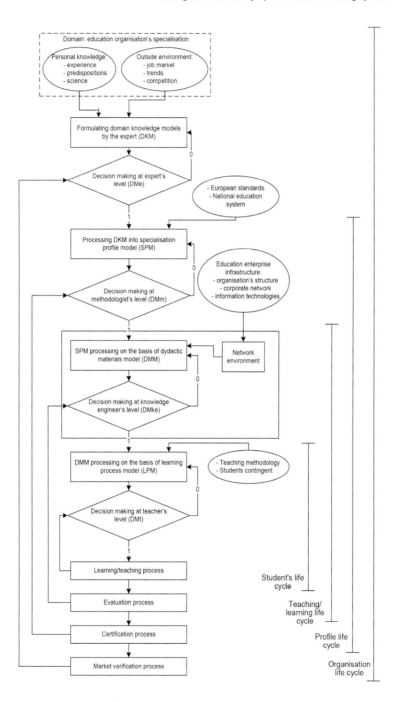

Figure 1.5 Functional scheme of the open distance learning system (OSDL).

where *Pr* are production processes, *R* are roles, *Ac* are activities, *Kd* is the domain knowledge.

The DKM is formulated by an expert on the basis of arising market demands for new processes, technologies and enterprise organisation forms and for establishing new roles and redefining tasks for the domain specialists. The expert focuses on the domain that the given organisation specialises in.

2) *Specialisation profile model (SPM)* relates to the given specialisation syllabus management subsystem (LCMS). The structure of this model can be described as follows:

$$SPM = \{DKM, Ks, Sk, Ab\}, \tag{1.2}$$

where *Ks* is the specialist's theoretical knowledge, *Sk* are practical skills and *Ab* are abilities.

The MPS is formulated by a methodologist (e.g. education officials) within the aims and structures of specialisations oriented on a given market area.

3) *Didactic materials model (DMM)* is formulated by a knowledge engineer on the basis of the specialisation profile and the objectives and structure of the learning subject. The DMM relates to the didactic material content management subsystem (LCMS). For a given subject, the DMM structure can be described with the following tuple:

$$DMM = \{SPM, S_y, G_{LO}^H, EC\} \tag{1.3}$$

where *Sy* are syllabus and learning objectives of a given subject, *G* is the hierarchical graph reflecting the subject's structure, *LO* are learning objects and *EC* are network environment constraints.

The knowledge engineer needs to have in mind the network environment constraints that are technical constraints of the learning space.

4) *Learning process model (LPM)* relates to the LMS. The model's structure can be described with the following tuple:

$$LPM = \{DMM, ISy, LE, LS, CP\}, \tag{1.4}$$

where *ISy* is the individual syllabus, *LE* are learning events, *LS* is the learning object sequence and *CP* are control points.

The LPM is formulated by a teacher, who bases it on the syllabus and learning of a given subject, the content of didactic materials and the evaluation of the initial knowledge of a given contingent of students.

1.2.4.2 Management outlines

The structure of a typical management outline (MO) in the ODLS is shown in Fig. 1.6. The MO corresponds to each management cycle. The main goal of the outline is managing the process of adapting the descriptive knowledge to the normative one within the outline.

The decision-maker of level *i* can be described with the tuple $\overline{C_i}$:

$$\overline{C_i} = \{MA_i, IV_i, F_i, PR_i, DF_i, MC_i\}, \tag{1.5}$$

where MA_i is the management activity coming from the top level; IV_i are the constraints (interfering variables) coming from the exterior environment; F_i is the feedback coming from the control process; PR_i the production rule, according to which the management activity for the bottom level is being formulated; $DF_i = MA_{i+1}$ is the decision function, according to which the content of management action entering the inferior level is being estimated; and MC_i is the management cycle at level *i*, meaning the time interval limiting the management action MA_i.

The diagram shown in Fig. 1.6 can be considered as a MO with feedback F_i. The superior level decision-maker is the subject of management. The inferior level decision-maker together with the learning process are the objects of management. The central system-making element of the MO is the decision-making circuit that compares the normative knowledge model (NM) with the descriptive knowledge one (DK). The normative knowledge model represents valid, proper behaviour rules, while the descriptive knowledge model impartially and objectively describes reality (Broens and de Vries, 2003) (Zaikin *et al.*, 2006). The normative knowledge model is formulated by the system's decision-maker on the basis of the management activity entering from the superior level. The descriptive knowledge model is specified by the inferior-level decision-maker.

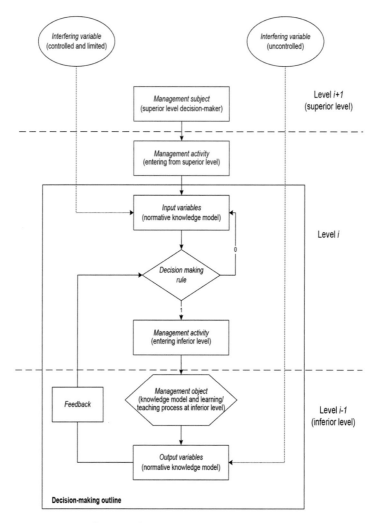

Figure 1.6 Open distance learning system (ODLS) management outline structure.

As was shown in (Zaikin *et al.*, 2006), the knowledge representation model most suitable for the researched system is a hierarchical concept graph G^P. If methodology and algorithms from (Zaikin *et al.*, 2006) are applied, knowledge models *NK* and *DK* can be presented in the form of hierarchical graphs G_{NK}^P and G_{DK}^P. With this approach, decision rule PR_i, on the basis of which decisions are made, can be described with the Kronecker symbol:

$$PR_i = \begin{cases} 1, & \text{if } G_{NK}^P \supseteq G_{DK}^P \text{ and } HD_i \leq T_i \leq HD_{i+1} \\ 0, & \text{otherwise,} \end{cases} \tag{1.6}$$

where G_{NK}^P is a hierarchical graph of the normative knowledge model, G_{DK}^P is a hierarchical graph of the descriptive knowledge model, HD_i is a decision time horizon at level i, HD_i is a decision time horizon at the inferior level $i+1$.

Each production rule PR_i consists of two conditions:

a) Normative knowledge graph G_{NK}^P covers the descriptive knowledge graph G_{DK}^P.

$$G_{NK}^P \supseteq G_{DK}^P \tag{1.7}$$

b) Decision-making period T_i is longer than decision horizon HD_i and shorter than decision horizon HD_{i+1}.

$$HD_i \leq T_i \leq HD_{i+1} \tag{1.8}$$

As can be seen in Fig. 1.6, when both conditions are fulfilled the decision-maker formulates a management action for the inferior level $(i-1)$ on the basis of the decision function DF_i. In this case, the decision function DF_i changes the normative knowledge model NK_{i+1} of the inferior level.

Should one of the described conditions not be fulfilled, the decision-maker stays at level i to develop and modify his/her own normative knowledge model NK_i.

1.2.4.3 Content of the management system

The approach described before is applied in every cycle of education organisation management. Let us define the content of each management cycle presented in Fig. 1.6.

Organisation life cycle A subsystem of education organisation strategic management aiming at maintaining a high position of the organisation's graduates at the job market. The expert's decision model has the following form:

$$DM_e = \{MA_e, IV_e, F_e, PR_e, DF_e, MC_e\}, \tag{1.9}$$

where $MA_e = \emptyset$ is the management activity, IV_e is the market demand for the specialisation, SE_e is the periodical graduates control in order to estimate their satisfaction in reference to market needs, DF_e is the decision function (creating new specialisation, modifying an existing one), LC_e is the organisation life cycle and PR_e is the expert's production rule in the following form:

$$PR_m = \begin{cases} 1, & \text{if } G^P_{NKm} \supseteq G^P_{DKc} \text{ and } HD_m \leq T \leq HD_{ke} \\ 0, & \text{otherwise,} \end{cases} \tag{1.10}$$

where G^P_{NKm} is a hierarchical graph of normative knowledge at the methodologist's level (knowledge needed during the specialisation's profile creation), G^P_{DKc} is a hierarchical graph of the student's descriptive knowledge from the certification process and HD_m is the methodologist's decision horizon and HD_{ke} is the knowledge engineer's decision horizon.

Teaching/learning life cycle A subsystem providing an intelligent and network space of the learning/teaching system (effective usage of a network environment and developing or adapting knowledge repository to the student's profile and students' contingent). The knowledge engineer's decision model has the following form:

$$DM_{ke} = \{MA_{ke}, IV_{ke}, F_{ke}, PR_{ke}, DF_{ke}, LC_{ke}\}, \tag{1.11}$$

where $MA_{ke} = DF_m$ is the methodologist's management activity (changes in the specialisation profile, changes in didactic materials); IV_{ke} are the organisational structure, corporate network, software and technical resources, F_{ke} is the statistical data from the student's evaluation process, DF_{ke} is the knowledge engineer decision function (changes in didactic materials, changes in learning/teaching method-ology), LC_{ke} is the student's life cycle and PR_{ke} is the knowledge engineer's production rule in the following form:

$$PR_{ke} = \begin{cases} 1, & \text{if } G^P_{NKd} \supseteq G^P_{DKs} \text{ and } HD_{ke} \leq T \leq HD_t \\ 0, & \text{otherwise,} \end{cases} \tag{1.12}$$

where G^P_{NKd} is a hierarchical graph of normative knowledge at the didactic level (knowledge included in didactic materials), G^P_{DKs} is a hierarchical graph of the student's descriptive knowledge within the subject of learning, HD_{ke} is the knowledge engineer's decision horizon and HD_t is the teacher's decision horizon.

Student life cycle A subsystem that allows following and monitoring administrative correctness of the learning process, and evaluating the competence-gaining process with the given knowledge model and teaching/learning system. The teacher's decision model has the following form:

$$DM_t = \{MA_t, IV_t, F_t, PR_t, DF_t, LC_t\}, \tag{1.13}$$

where $MA_t = DF_{ke}$ is the management activity of the knowledge engineer (changes in didactic materials, changes teaching/learning methodology), IV_t is the contingent of students, F_t are the students' grades during the learning process, DF_t is the teacher's decision function (forming students groups), LC_t is the subject's learning cycle and PR_t is the teacher's production rule in the following form:

$$PR_t = \begin{cases} 1, & \text{if } G^P_{NKt} \supseteq G^P_{DKe} \text{ and } O_e \leq T \leq HD_{ke} \\ 0, & \text{otherwise,} \end{cases} \tag{1.14}$$

where G^P_{NKt} is a hierarchical graph of normative knowledge at the teacher's level, G^P_{DKe} is a hierarchical graph of the student's descriptive knowledge from the examining process, O_e is the examining period and HD_{ke} is the knowledge engineer's decision horizon.

1.2.5 Conclusion

The presented approach is based on a commonly acccpted and verified theory, and that stands for its high reliability. The multi-level, hierarchical systems theory is used often nowadays (examples can be found in (Gómez *et al.*, 2001) and (Miklashevich and Barkaline, 2005)). The reason for that is the fact that the world's economy is based on more and more complex systems (e.g. EU, companies and firms with a corporate character).

By identifying management cycles we have acquired the ability to distinguish knowledge layers that are the basis of decision-making. Usage of existing knowledge representation methods, for example conceptual graphs or topic maps, will enable building a best-practice database.

The next step in the research being carried out by the authors will be extending the proposed apparatus with the ability of comparing the semantic content of each knowledge model. In (Zaikin *et al.*, 2006) a method of comparing single concepts by taking into consideration their capacity and depth in different contexts was presented. New tasks standing before knowledge models require developing mechanisms that allow comparing descriptive and normative knowledge models of the same knowledge domain.

1.3 Student Life Cycle

1.3.1 Proposition of the Student Life Cycle

Our approach to the student life cycle is based on the competency indicator objective. It may used for validation and verification of the entire knowledge-based learning process (KBLP). We consider competency as a skill of theoretical knowledge and practical knowledge use in real situations (applications). The competency range is set on the basis of fundamental knowledge being the requisite base for procedural (specialised) knowledge. The fundamental knowledge is an abstract type of knowledge, enabling the object effective reasoning in any technical domain. An object's competency (of engineer grade, for example) depends on abilities of merging both the fundamental knowledge and the know-how.

The scheme proposed in Fig. 1.7 models a student's intellectual progress as knowledge-based process. The scheme represents major layers in KBLP management. Five levels have been isolated to enable controlling and steerage of knowledge transfer and acquisition. Knowledge characteristic varies, in accordance with the following stages of the learning process. The knowledge is not considered as a monolithic model; it is formed from a consistent set of models addressing and encompassing several education activity objectives.

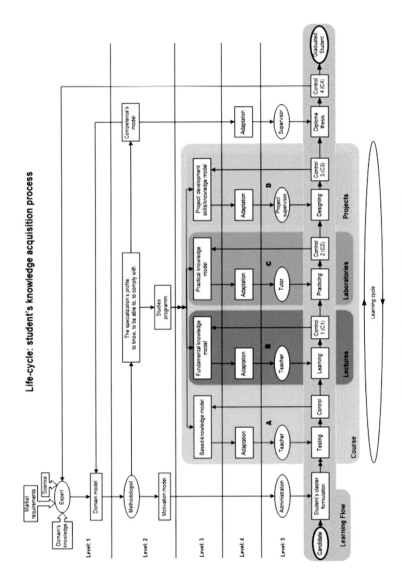

Figure 1.7 Life cycle student's knowledge acquisition process.

Several actors are involved in the presented student life cycle. These are professionals from certain domains working at the particular organisation. We are particularly interested in the actors' influence on the process of forming and consolidation of a student's knowledge in the context of a given scheme.

The approach is characterised by the following entities:

Definition of the actors:

Expert: Single, or group of professors responsible for the university's public appearance and position. The highest competency indicator is represented by the expert.

Methodologist: Is responsible for adaptation of the domain model developed by the expert into limitations and constraints of the organisation (this denotes learning process internal quality). The task is defined as follows:

(i) On the one hand, formulation of the course specialisation's profile, transposed into real planning of lectures, laboratory and project courses

(ii) On the other hand, development of a motivation model (MM), transposed into a real education offer partial to the university

Administration: Is responsible for the MM analysis (education offer) and, based on this construct, forms the students' contingent. Additionally, it also plays a key role in the deployment of knowledge requirements meeting the organisation's processes.

Teacher, tutor, supervisor: Are directly responsible for the courses. They adapt the education situation to a group of students, depending on their domain knowledge and pedagogical knowledge. Specifically, a teacher concentrates on lectures – theoretical knowledge; a *tutor* on laboratory courses – procedural knowledge acquisition; and a supervisor on project courses and the final degree certificate.

Let us specify the levels of the given scheme including the models and actors. Each level concretisation provides each actor's function definition.

1.3.1.1 Multi-levelled structure of the learning process

The multi-levelled structure of the learning process is represented in Fig. 1.7.

Level 1 – *expert's level: analysis and innovation support within the learning process*

The expert by means of analysis of the labour market situation (within region), recognition and identification of the current state of science and domain knowledge, develops a DM. Its main objective is to represent the required knowledge at the requested time interval. The model's validity and timeliness are interpreted as the difference between knowledge gain inherent to the domain and the knowledge model which was enclosed in the student's profile. This concludes a support in relation between the labour market, professional scientific knowledge and the university's intellectual responsiveness. The analysis of this set resolves into adaptation and continuous improvement of the domain knowledge model, producing innovation in each cycle of the learning process.

Level 2 – *strategy of the learning process planning*

This level enables initial setting of learning process objectives. The DM is considered as a reference model of the student's knowledge and determines adaptation parameters of didactic materials for the student. The Motivation Model (MM) is a result of the methodologist's analysis of the Domain Model (DM). The MM indicates which students (specialists) will be prepared by the organisation. Feedback with Level 1 ensures that the obtained edification shall secure the students' employment in the future.

On the strategic level, a particular organisation is adapted to market requirements. The adaptation process is a result of DM interpretation in the context of the organisation's practical constraints, mission and strategic goals. This revolves into the specialisation course profile. Next, the profile is transposed into the subject list, which creates both the education process perspective (horizon) and the competence model (CM). The education process perspective is a background for a study plan at the organisation (university). It is a result of requirements addressing particular educative activities – modification and innovation within lectures, exercises, laboratory and project courses. This may be specified in the following sequence: course development, course modification and knowledge range alteration within the course type. Furthermore, other subjects are included, such as mathematics, which are an upshot of the education organisation's mission. The CM is a result of the total knowledge required and specifies the requirements level of the final diploma project course.

Level 3 – *Didactic materials content formulated available online*
At this level, using outlines and settings from Level 2, the subject board content is specified for each course included in the specialisation profile. Each course is pervaded with knowledge according to the strategy prior to DM analysis. The subject board structure and representation mode in distance learning and traditional learning vary significantly, depending on SCORM and the knowledge repository.

Level 4 – *adaptation operation* – *preparation of the students' contingent*
Adaptation operation provides an adaptation possibility of the reference knowledge form comprising the models from Level 3 into a real education situation. At this level, students are identified (considered as a group). The situation is altered by creation of personalised selection (perspective on) of the knowledge on the basis of the teacher's/tutor's/supervisor's knowledge. The knowledge is not cancelled, but the moment of its transfer and application order is positioned on a timeline.

Level 5 – *Implementation and application of the KBLP and didactic materials preparation*
This level represents the implementation and application process of the composed learning process (KBLP), resulting in making the didactic materials accessible. All the actors work in the online consulting mode.

At Levels 3, 4 and 5 the following components have been integrated: students' contingent formulation process and feedback loops (*A, B, C, D*) characterising the factual learning process within the contingent. The proposed approach includes divergence of the knowledge processed according to the subject and its specifics. Therefore, four learning process loops (feedback) have been outlined. Each of the loops is devoted to a different purpose and distinct characteristics. These are the following:

A Student's base knowledge analysis: The teacher validates the student's level of qualifications for a given course. The results become a vital parameter and criterion for the adaptation operation, which is used sequentially through the learning flow process. Testing process provides the student's knowledge range measurement within a particular domain. This is relevant to set the student's competence degree within the domain. Moreover, the student's knowledge quality may be investigated, for example, by means of the concept's depth analysis and identification.

B Stage of fundamental knowledge absorbance (learning), represented by lecture course: The teacher transfers abstract knowledge of the domain of discourse, enabling a student's mastering of abilities of abstract concept handling and reasoning within the given knowledge system.

C Procedural knowledge edification: Teaching is focused on software use and comprehension of particular computer-supported simulation areas (environments) functioning (comprehension of the applied metaphors). The tutor supports the student to transform his/her fundamental knowledge into an actual computer programme or events required for performing the simulation execution.

D Application of absorbed knowledge in real situation: The aim of this stage is to apply acquired knowledge (stages *B, C*) in a concrete real event. The student is expected to classify the given task categories, which are to be solved, efficiently and skilfully refer to his/her own cognitive schemes and apply the appropriate tool. The process is finished by obtaining results analysis and conclusions production.

Each of the loops refers to three levels: development or choice of the LO (*level 3*), adaptation of the factual LO (*level 4*) and testing and validation of the test results (*level 5*).

The final mechanism of the entire learning process validation is a process of diploma formation and development. A diploma thesis consists of the knowledge acquired by the student during the entire learning process execution. Correspondingly, an actual DM form has influence on the thesis shape. This provides an opportunity for correction of conceivable deficiencies in order to achieve the major goal: the student possesses knowledge, strongly related to and meeting the market requirements.

1.3.1.2 The quality issue

The quality of the learning process may be considered as efficiency of knowledge transfer into the student's mind. The decisive quality indicators are generated by controlling components ($C1$, $C2$, $C3$, $C4$). Efficiency in identification of the parameters influences the adaptation process, which defines a form of knowledge transfer into a particular situation, specific to the contingent of students. In case the adaptation operation fails, all the knowledge transfer may become inefficient.

The quality of university functioning is defined by efficiency in formulating and implementing DMs through the study duration. At the expert level, when a university: dynamically and flexibly alters its study plan minding, the environment requirements (e.g. the labour market), economical calculation and pedagogical conditions, the graduates succeed on the labour market, affecting the quality of the university.

1.3.2 Conclusion

The student life cycle should include and integrate two processes, learning process, proposed by Universitat oberta de Catalunya (UOC), and the proposed approach of the KBLP. Integration offers abandoning constrains and cultural differences between students of Europe. Apart from the development of a knowledge repository, the KBLP supports adaptation and individual classification of the various education situations (classification prior to student characteristics). Each student is processed separately because subsequent knowledge models are formed (A, B, C, D) specific to the student's characteristics (obtained at the testing stage).

The proposed model of the student life cycle offers precise clarification at which stage and what didactic materials are used. The didactic materials should be in a modular form to make adaptation operation probable, apparent and effective, which is considered inefficient and tedious, based on hypertext-driven materials.

The proposed materials offer solutions on organisational aspects of an education institution. Efficient division and assignation of functions to workers is expected. Additionally, the model brings a certain notion about the required worker's competencies.

Didactic materials are, in major, supported by the hypertext mechanism. The materials may only be verified in the context of computer technology application, not from the cognitive point of view. The proposed model indicates possibilities of a solution of this problem.

Final conclusions:

- To increase the clarity of the proposed model, a thesaurus should be developed, providing clear, interchangeable definitions of concepts occurring in the model.

- The model may be interpreted on a timeline, beginning from the left, and/or from the top of the scheme.

- The model includes consistent and closed feedback as an integral management system of education organisation and the learning process.

Chapter 2

Methods of Modelling of Competence

2.1 The Meaning of the Concept of 'Competence'

Many experts in such fields as sociology, pedagogy, philosophy, psychology and economics have tried to define the notion of competence. Their efforts have been shaped by educational, cultural and linguistic contexts. Romainville (Romainville, 1996) indicates that the French term 'compétence' was originally used in the context of vocational training, referring to the ability to perform a given task. However, in recent decades, it has found itself in general education, where it often refers to the 'possibility' or 'potential' for acting efficiently in a given situation. What matters is not so much knowledge itself as the ability to apply. According to (Perrenoud, 1997), competence means enabling individuals to mobilise, apply and integrate acquired knowledge in complex, diverse and unpredictable situations. He defines competence as 'the ability to effectively act on a number of specific situations, a skill based on knowledge, but unlimited by it'. In a document published by the Organisation for Economic Co-operation and Development (Rychen and Salganik, 2001), after analysing many definitions it was found that in many of the disciplines, competence is interpreted as

Open Distance Learning: Fundamentals, Developments, and Modelling
Oleg Zaikin
Copyright © 2023 Jenny Stanford Publishing Pte. Ltd.
ISBN 978-981-4877-55-8 (Hardcover), 978-1-003-13261-5 (eBook)
www.jennystanford.com

a roughly specialised system of abilities, skills or agilities that are necessary or sufficient to achieve a particular goal. At the symposium the Council of Europe on competencies (Hutmacher, 1997) proposed that the term 'competence' or 'competencies' is understood as 'general ability (possible) based on knowledge, experience, values and dispositions acquired as a result of educational activities'.

Many authors in the field of management, among others (Cardy and Selvarajan, 2006), (Partington *et al.*, 2005), (Crawford, 2005), show the work of (Boyatzis, 1982) entitled 'The Competent Manager' as that which introduced and popularised the concept of competence to the literature in the field of management. In this work, Boyatzis defines competence as 'a person's character, defined by such characteristics as motivation and skills expressed in his/her image, social role or actively used knowledge'. Such a broad definition could indicate that competence is any distinctive personality trait, but Boyatzis narrowed the concept of competence only to the individual characteristics manifested in the context of the performance by the unit of work performed by it. In addition, Boyatzis distinguished competence of the tasks and functions assigned to individuals in organisations, indicating that they are what people bring to their jobs and not what they perform as part of their duties.

(Woodruffe, 1992) defines competence as 'a set of behaviour patterns needed for the proper performance of duties or functions'. This definition has a very simple but important characteristic. Namely, the notion of competence may contain knowledge, skills and abilities, but also can go beyond these traditional characteristics, in particular considering the motivation and willingness to perform jobs.

Thus, competencies are seen as characteristics of a person associated with the successful execution of a task. These characteristics should be reflected in the observed patterns of behaviour that have a positive impact on the work carried out. Among those observed behaviours most often mentioned are knowledge, skills and abilities (Cardy and Selvarajan, 2006), but it should also be noted that besides them there are a number of personal features that could not be easily observed and identified and that may have a significant impact on the success made by person tasks. Hofrichter and Spencer (Hofrichter and Spencer Jr, 1996) define these characteristics by a term which loosely translated as a 'second bottom' of the employee characteristics or 'soft' skills ('below the waterline'). For example, such features as a professed system of values and personality of the employee can affect a similar degree of success in the execution of tasks such as his/her technical skills. Therefore, in the case of identification and competency requirements, it is necessary to take into account not only the competence directly related to such performance but also various kinds of personality features having a positive impact on the success of the allocated tasks. Furthermore, any models utilised in the analysis of competence should allow mapping of these two types of competence.

The concept of competence model in the literature is often used interchangeably with the term 'competence'. (Mansfield, 1996) defines competence model as a detailed behavioural description of the characteristics of an employee that are necessary to be effective. The form of this description can vary and take the form of av erbal description or a mathematical model. A competence model can then be used to present a set of competencies associated with a given task, position or role in the organisation.

As mentioned before, the concept has been defined in the context of management science. According to Cardy and Selverajan (Cardy and Selvarajan, 2006) competencies can be a major source of competitive advantage, and even the best-prepared strategies cannot be successfully implemented and realised without appropriate competent stuff.

The concept of competence may be viewed differently within the organisation. From the view of strategic management, competence of the entire organisation is defined as a combination of its resources and capabilities (Hitt *et al.*, 2005). The problem of competence also developed as mainly relevant to *human resource management (HRM)* and started to play an important role in various areas of management, such as recruitment, selection, training, career development and the reward system (Partington *et al.*, 2005). From the point of view of HRM, competencies are seen as skills of employees, which define the characteristics of an employee that have to be performed and the requirements of work at a given position (Cardy and Selvarajan, 2006). The main issue analysed in HRM is to understand the fact that some employees perform their tasks better than others. An understanding of these mechanisms would allow more effective HRM techniques (Partington *et al.*, 2005).

A similar classification of competencies was presented by Turner and Crawford (Turner and Crawford, 1994), who shared personal and corporate ones. *Personal competencies* are permanently held by persons and are composed of features such as knowledge, skills, abilities, experience and personality. In contrast, *corporatecompetencies* are assigned to the entire organisation and are determined by existing inside the processes and structures that should not be changed when leaving the individual.

Since the competencies are intangible their observation and identification cannot be evaluated directly (Heywood *et al.*, 1992). For this reason, it is necessary to obtain some indirect evidence based on which the competence can be established. This observation led to the development of the two main approaches in the analysis of competencies:

1) *Presentation of competencies based on a set of attributes*: This approach focuses on the set of behavioural attributes that allow to build a competence model and tests the levels of these attributes in people whose competence is assessed. (Spencer and Spencer, 1993) identified five attributes that characterise competence. Two of them, the knowledge determined by the information held by the entity in a given context and the skills that are the ability to perform tasks manually or mentally, are relatively easy to formalise and evaluate. However, three others, related directly to the individual psychological profile of the entity, cannot be easily described and assessed. These three attributes are motivation, character and level of self-esteem.

2) *Presentation of competence as a demonstration of the efficiency in conducting activities related to a specific workstation*: This approach relies on the definition of typical activities performed at the workplace and related competencies. Then the person who performs work in this position is assessed for the presence and levels of these competencies. This approach was initiated in the U.K. in the process of creating a system of national training programs, such as the *National and Scottish Vocational Qualifications Management Charter Initiative (now acting as the Management Standard Centre)*. This approach has resulted in the development of standards of competence – systems

definition and assessment of competence for various branches of economic activity and existing in the typical workplace.

(Crawford, 2005) merged both approaches by proposing an integrated competence model allowing for assessment of competence on the basis of behavioural attributes such as efficiency of performance of tasks by the entity (Fig. 2.1). According to this model, competence is determined on the basis of skills and knowledge, which are treated as input. Personal competencies are regarded as attributes of the individual, defined and stable over time. By contrast, output competencies' level is determined by observing and in relation to the adopted standard of competence.

This review of competence definitions shows that its meaning is broad and can often be ambiguous. Its use in management/education requires the adoption of a single coherent definition. Finally, the International Organization for Standardization (ISO), in ISO 9000: 2005 'Quality Management Systems', provides in a concise and precise form the meaning of competence as 'demonstrated ability to apply knowledge and skills'.

3) *Competencies and qualifications*: In everyday language, the literature and unprofessional terms, the concept of 'competence' is often used alternatively with the term 'qualification'. The *PolishLanguageDictionary* defines qualification as 'training and abilities needed to perform some function or profession'. The concept of competence defines, in general, knowledge and skills owned by a person, while qualification is understood as holding of certain evidence of formal training or specific skills. The confirmation of formal qualifications may be a high school diploma or a university certificate, references and so on. In the same sense the concepts of competence and qualification are used in various national systems of vocational training. The concept of competence means there are different kinds of skills that an employee may have, while qualification is a certain level of skills attested by a diploma or certificate (Siciński, 2003).

In some sources the concepts of qualification and competence are different with regard to the skills that they relate. Qualifications are usually more associated with physical, manual or routine skills, while competence is more associated with intellectual activities (Siciński, 2003).

To summarise these considerations, competence is a general concept that defines the ability to perform different patterns of behaviour based on accumulated knowledge and experience, while qualifications relate to all kinds of formal evidence confirming possession by a person of specific knowledge and skills. Simply put, qualification is formal evidence of specific competencies.

Computer-aided management of human resources in a project requires the use of a formal model of competence, which enables one to quantify the usefulness of the research team to participate in the project. The previous definitions and the integrated model of the competence (Fig. 2.1) reflect only the nature of the competence and don't provide tools for quantitative analysis of the competence. In cases when an exact quantitative analysis of competence is required, it is necessary to rely on a model that will provide mathematical foundations and tools to carry this out. This model can precisely describe the competencies and their comparison, determine the cost of competence increase, determine the adequacy of the competence of the individual to the aim of the tasks and solve many other problems of a quantitative nature.

In the literature there are a number of mathematical models describing competencies. Most of them are based on the conception of (Yu, 1990) of habitual domains and the representation of competence as a set (Yu and Zhang, 1990; Shi and Yu, 1999) or a fuzzy set (Wang and Wang, 1998).

2.2 Standards of Competence

Today, many countries are working on competence standards. Setting standards refers to standard terminology used to describe competencies acquired through all stages of the learning process, and the creation of complex procedures for certification of acquired knowledge and skills. Standardisation assumes the competence acquired through not only formal education but also outside of it for various courses, training sessions, workshops and internships held before or during a person's working life. This allows to get a reliable

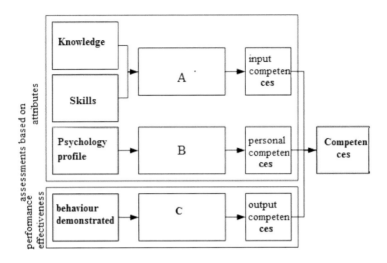

A-knowledge, comprehension, skills and abilities brought by a person to work

B-psychological profile of a person in the context of a completed task

C-ability to perform entrusted activities at a specified level

Figure 2.1 Integrated model to assess competence (Crawford, 2005).

image of the competence of the person regardless of where, how and at what stage of education or professional life the competence has been acquired.

In countries where standards of competence are being developed, competencies are defined as key measures having a fundamental nature and are identified as ways of acquiring, developing and obtaining relevant diplomas. The basis is a strong cooperation between government bodies, educational institutions, businesses and industry, as well the determination of general and more specific skills in such areas as physics, mathematics, computer science, languages, finance and management. In 2001, the European Commission entrusted a working group of national experts the task of defining the concept of 'key competence' and identifying those competencies that are recognised in all European Union (EU) member states (Smoczyńska, 2005).

This kind of standardisation allows the realisation of an EU project of a personal card of competencies, allowing any individual to present the skills that he/she obtained on the educational path. The purpose of this card is to develop tools that would allow for the gradual creation of common standards of competencies that exceed the vocational division. The European accreditation system of technical and vocational skills will come into force on the basis of cooperation. It will be to refer to high school education (development of software packages, qualification standards, the search for forms of financing and evaluating), branch unions, businesses and local chambers of commerce. In this action the social partners have to be included (Siciński, 2003).

Development of standardisation and certification systems of competence is essential in the context of methods to build project teams. The existence of standards of competence solves the problem of ambiguities in assessing individual competence, defining requirements of competence and building a formal mathematical model.

2.3 Competence Model for Project Management

In this part the problem of the selection of teams for the consortium implementing the project is analysed. In each project, besides the tasks of implementation of design work, there are also the tasks of project management. To determine the requirements of competence, the individual responsible for project management can use the existing standards for the area. The most commonly used and accepted standard is the Compendium of Knowledge for Project Management (Project Management Body of Knowledge [PMBOK]). It was published in 2004 by the Project Management Institute (PMI) (PMI, 2004) – a consortium of experts and practitioners around the world that develops the methods of project management. PMBOK is a set of standards and best practices for all identified processes oriented to project management. In the United States, PMBOK has been approved by the American Institute of Standards (ANSI) as an official standard of project management.

PMBOK defines areas of expertise and project management processes (Fig. 2.2) that may be treated as competencies required to manage specific areas of a project.

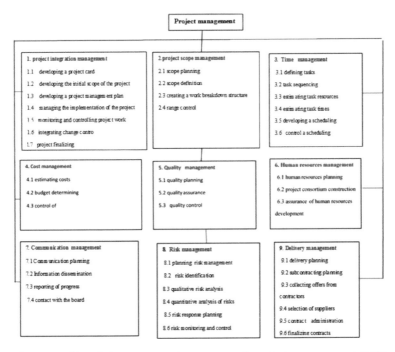

Figure 2.2 Knowledge areas and processes of project management (PMI, 2004).

All project management processes are classified into five groups of processes: initiating, planning, implementing, monitoring and control and concluding (Table 2.1). The presented method of selection of teams to the project consortium aims at such a choice of the partners when the consortium is prepared for the implementation of the project from the beginning of grant funding. At this point, according to most research funding programs, there must be a detailed plan for the project and the composition of the consortium must be fixed. Therefore to define the competencies required to manage the project from the granting of funding, a group of processes related to the execution, monitoring, control and finalisation of the project can be used. On the basis of these group processes, 44 competencies can be defined necessary to describe the requirements for the team (most often it is appointed by coordinator of the project) dealing with the project management.

Table 2.1 presents the processes defined by the standard *PMBOK*. The skills of management may be necessary for the scientific coordinator of the research project. These processes can be defined as competencies and used to describe the requirements of coordinator. By comparison of the competencies of the potential coordinator with the competencies required for the project, the usefulness of playing the role of coordinator can be evaluated.

2.4 Competence-Based Approach: Representation of the Structure and Range of Competence

The situation of higher education institutions in the world is changing. This is caused by many factors, the most serious of which are:

- The dynamic development of new technologies
- The need for continuous education
- Adjusting education profiles and study programs to the market requirements
- The concept of competence-based learning, strongly supported by the labour market and the European Social Fund

The competence-based learning process under new conditions of education is represented in Fig. 2.3. There are two kinds of markets, educational and labour. On the one hand an educational institute develops the learning profiles of specialties, according to which the institute provides training.

On the other hand, the labour market forms a request for a certain number of workplaces with a certain qualification. Both markets, educational and labour, are connected to each other through competencies. On the basis of these competencies, the labor market forms requirements for the educational market through the *required competencies*. In turn, the educational market forms proposals for the labour market through *guaranteed competencies*.

Table 2.1 Processes of project management (PMI, 2004)

Area of knowledge	Group of the initialisation processes	Group of the planning processes	Group of executive processes	Group of monitoring and control processes	Group of finalising processes
1. Management of project integration	1.1 Development of project chart 1.2 Development of initial scope of the project	1.3 Development of project management plan 1.3.1 Determination of tasks 1.3.2 Sequencing of tasks 1.3.3 Estimation of resources of tasks 1.3.4 Estimation of time of tasks 1.3.5 Development of a schedule	1.4 Leading and managing a project execution	1.5 Monitcring and controlling the work 1.6 Integrated control of changes	1.7 Finalising of the project
2. Management of project scope		2.1 The scope defining 2.2 The scope planning 2.3 Development of the work structure		2.4 Verifying of scope 2.5 Control of scope	
3. Time management		3.1 Determination of tasks 3.2 Sequencing of tasks 3.3 Estimation of resources of tasks 3.4 Estimation of time of tasks 3.5 Development of a schedule		3.6 Control of schedule	
4. Cost Management		4.1 Estimation of the costs 4.2 Budgeting		4.3 Control of costs	
5. Quality management		5.1 Planning of the costs	5.2 Ensuring of the quality	5.3 Controlling quality	

(Continued)

Table 2.1 – *(Continued)*

Area of knowledge	Group of the initialisation processes	Group of the planning processes	Group of executive processes	Group of monitoring and control processes	Group of finalising processes
6. Human resource management		6.1 Planning of human resources	6.2 Construction of the project team 6.3 Ensuring the development of the project team	6.4 Project team management	
7. Management of communications		7.1 Planning of communication	7.2 Dissemination of information	7.3 Reporting of progress 7.4 Contact management with shareholders	
8. Risk management		8.1 Planning of the risk management 8.2 Identifying of risk 8.3 Qualitative risk analysis 8.4 Quantitative risk analysis 8.5 Planning of the risk actions		8.6 Monitoring and Control of risk	
9. Supply management		9.1 Planning of purchasing and supply 9.2 Planning subcontracting	9.3 Collecting offers from suppliers and subcontractors 9.4 Selection of suppliers and subcontractors	9.5 Contracts administration	9.6 Finalising of contracts

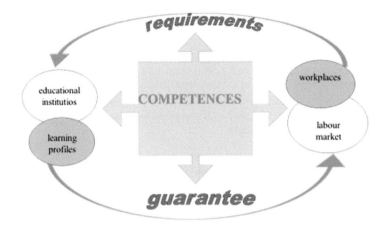

Figure 2.3 Competence-based learning process as a new way of education.

Competence is a result of a learning process in open distance learning conditions and can be defined as the following triplet:

$$K = W_t, W_p, W_{pr}, \tag{2.1}$$

where

W_t is theoretical knowledge,

W_p is procedural knowledge, and

W_{pr} is project knowledge.

This definition is the basis of ontological graph design. An *ontological graph* reflects the teacher's view on realisation of a given subject for a given speciality in ag iven area. The graph can be divided into certain portions of competence, for example groups of concepts connected by relations. These groups identify theoretical, procedural and project knowledge entering a given portion of competence. The definition of the competence is represented in Fig. 2.4.

Created in such a way, the graph is a base for repository formation as didactical material of a given content of theoretical, procedural and project knowledge. The portion of competence is a connected subgraph of the ontology graph, formed for each subject (course).

Competence is a collection of knowledge, attitude, skills and appropriate experience required to fulfil a given function

Competence is the ability to use theoretical knowledge to solve practical tasks and the ability to interpret the results in the terms of the used theory .
The competence learning process is interpreted as the ability
- to use theoretical knowledge
- to solve a practical task
- to interpret results in the frames of the used theory

Competence K can be described:
$$K=\{W_t, W_p, W_{pr}\}, \text{ where}$$
W_t – *theoretical knowledge,*
W_p – *procedural knowledge ,*
W_{pr} – *project knowledge.*

Figure 2.4 Definition of competence as a triplet of theoretical, procedural and project knowledge.

The preparation of the ontology graph is the result of integration of two elements. On the one hand it is a mental operation of a person who, due to the field of objects and the nature of the subject, determines the semantic relationships between concepts and the hierarchy of concepts. On the other hand, the preparation of graph 'needs' an implementation environment. For this purpose the tools available are, for example the program Protégé.

The prepared graph containing a set of concepts and relationships between them has a tree structure. The nodes correspond to the three types of knowledge. The development of ontology is done on the basis of adding a new class of concepts in a graph or single copies. Nodes are assigned appropriate teaching material. Analysis of the contents of teaching materials assigned to the vertices of the graph and placed in the repository is the basis for the development of competencies. The development of the repository takes place not only on the basis of the materials the teacher but also on the basis of the tasks of students. From the point of view of a student placed in the repository, tasks can be the basis for the development of e-portfolios

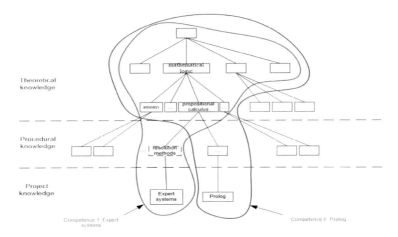

Figure 2.5 Representation of the structure and range of competence.

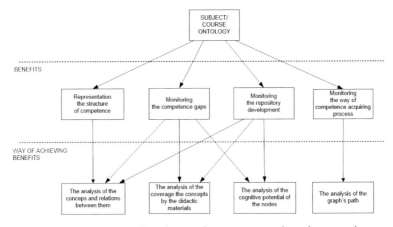

Figure 2.6 Main benefits of using the competence-based approach.

of students and shaping of a personalised career path. In turn, the position of the teacher and focus on a new model of education based on competence requires that the student turn in the creative process of repository development.

Competence is a collection of knowledge, attitude, skills and appropriate experience required to fulfill a given function. The learning competence-based process is interpreted as the ability to use theoretical knowledge to solve a practical task and the ability to interpret results in the frames of the used theory.

Ontology as a method of knowledge representation through specification of concepts and relationships between them gives the possibility to conceptualise a given knowledge domain. It is possible to specify concepts describing theoretical, procedural and project knowledge. It is possible to create a hierarchy of these concepts and to show the sequence between them (Fig. 2.5).

The main benefits of using the competence-based approach are represented in Fig. 2.6.

Chapter 3

Team Project Process Oriented on Acquiring Competence

3.1 New Educational Model Developed for Acquiring Project Team Competence

> 'Tell me and I will forget
> Show me and I remember
> Involve me and I will learn'
>
> —Benjamin Franklin, 1706–1790

Defining, developing and implementing a new educational model based on competence have been some of the most important challenges in recent years, and this has led to major changes in the methods and techniques of education.

3.1.1 Engineering in the Global World

One of the most important tasks of the Warsaw School of Computer Science (WSCS) and the Odense Technology University (OTU) in

Open Distance Learning: Fundamentals, Developments, and Modelling
Oleg Zaikin
Copyright ©2023 Jenny Stanford Publishing Pte. Ltd.
ISBN 978-981-4877-55-8 (Hardcover), 978-1-003-13261-5 (eBook)
www.jennystanford.com

recent years has been the definition of a new educational philosophy, as well as its development and implementation, which has led to large changes in educational methods and philosophy.

The new education model is the result of major changes in the contingent of students and staff of the WSCS and OTU. It is a set of values on which teaching and learning activities are based.

Among all the reasons for the emergence of the new education model, two most important must be distinguished:

- The need to increase the globalised labour market in basic (first rate) technical knowledge as well as in social and personal competence, intercultural and linguistic qualifications and adaptability of teaching methodologies and learning abilities
- The need to increase students' independent education and greater suitability in conducting the educational process

Nowadays, a traditional educated engineer with high technical skills meets with strong competition all over the world, because engineering work moves to countries with more competitive earnings. In addition, the requirement of an extended profile needs engineering education to evolve from traditional educational methods with an emphasis on technical content to a new educational culture in which a student combines a strong professional profile with a wide set of personal skills.

To ensure these requirements, the WSCS and OTU have defined a new student education profile in the form of the following skills and abilities:

- *Intellectual competencies*, that is, the ability to analyse, synthesise and evaluate
- *Personal competencies*, that is, the ability to communicate, work in a team and organisation, self-control and self-regulation in the team
- *Practical and project competencies*, that is, the solution of complex and inter-disciplinary projects in the context of the real world

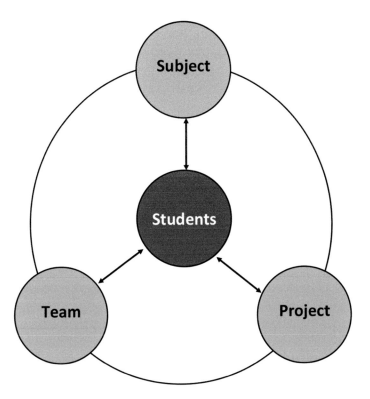

Figure 3.1 Students meet and collaborate with three basic elements in an educational environment.

The new student profile has evoked a new approach to education, the result of which is the implementation of a student-oriented learning environment that includes three basic elements (Fig. 3.1):

(1) Subjects structured more in lessons than in lectures
(2) Projects that were made by a team of students and a teacher as a co ordinator, often carried out for specific companies with the addition of real perspectives
(3) Students within the team performing specific roles

All activities in a specific environment are based on four basic principles that must be implemented in the functioning of the education model:

- Motivational environment
- Student activity
- Teamwork
- Structured knowledge

Let's consider these components of the education model.

(1) *Motivational environment* :

The basic motivational factor in the new model of education is the student's balanced choice of work on the subject in the right context. The choice of work determines the environment in which the student is motivated to be active and responsible for work and studies. In addition, we set our expectations, and so how the student performs his/her duties from the first day at the university. This approach increases the student's responsibility for the entire educational process, which gives the student a sense of independence.

Example: The OTU provides each student group with its own room for the duration of the project.

(2) *Student activity* :

Students are activated by choosing a specific project and re-placing traditional lectures with other forms of active teaching, such as experiments, group work and student presentations. To create conditions for the use of different teaching techniques, we exchange two hours of lectures for four hours of active classes.

The motivation to work with different teaching techniques is illustrated by the results of the 'Bethel Maine USA' National Education Laboratory project. These results are presented in Table 3.1.

As we can see in the table, different ways of transferring knowledge have different long-term effects on students, for example students with only 5% of lecture material after 24 hours but 90% of knowledge when they teach other students. In the educational model, we will use different teaching techniques for the best perception of knowledge by students.

(3) *Team work* :

The knowledge engineer plays a central role in the working

Table 3.1 Rate of effectiveness of retention

Lectures	5%
Mastering the teaching material	15%
Audio-video educational materials	20%
Demonstration technique	30%
Discussion groups	50%
Practical activities	75%
Teaching other people	90%

process and often encounters problems that need not just technical knowledge. Working in a team needs personal skills, too, to improve teamwork and achieve better results. In any team and especially in teams from various professions, countries and cultures that are becoming more common, there are difficulties in taking decisions that can lead to conflicts. To resolve such calls, we introduce different teaching strategies that are introduced in real working situations, together with various tools to solve non-technical problems, for example cooperative conflicts and group process management. Students' understanding of the sense of teamwork as current engineering gives more opportunities for trust and experience at the first stages of teamwork, which in the future can be used in a professional student career.

(4) *Structured knowledge:*
   Students face the problem of presenting, interpreting and processing knowledge in every subject from the first day of teaching at the university. Knowledge of each subject is not isolated and needed when it is needed in a given context.

In WSCS models, teaching is a combination of action and reflection, supported by knowledge structuring, experimentation and student experience. For full understanding, mastered knowledge must be used to solve a practical task. To this end, tests must be created in every subject to use new knowledge in a team project. For the content of the subject to be fragmentary and better understood, the concepts must be organised around specific topics. Students meet the same topic in different subjects, but in different aspects and perspectives.

For example, the semester topic 'System Integration' includes several different subjects, such as 'Programming' and 'Data Communication'. Both subjects are part of a team project. The project team consists of five to seven students who can integrate, apply and increase knowledge in solving a real problem in a real context in the real environment of an existing company before students receive a legitimate consideration of the subject and problem.

From the professor's point of view, an important factor is the transition of students from the lecture hall into a specific production environment and students become direct participants in the design/production process in a real environment. This means that a base is created for the motivation of all participants in the teaching process, increasing the speed of the process of students learning the knowledge.

The role of the teacher (tutor, professor) also changes significantly. The teacher plays a more important and diverse role in the interaction of students in the research and design process. In addition, the professor also performs the functions of:

- Supervisor: support and supervision of student work
- Consultant: checking and repairing tests

In the team project model, each professor is a participant in the team in which he/she creates a well-balanced unity. The team project motivates professors to pass on the student's experience and as a result he/she makes a great contribution to the team's shared knowledge.

3.2 Scenario of the Learning Situation

3.2.1 Orientation of the ODL Process on the Active Cooperation of Students and the Teacher

The implementation of the open distance learning system (ODLS) provides for a change in the entire paradigm of organisation of the learning process and thus a change in the role and relations between all participants of the learning process, while maintaining the status quo regarding the traditional mission of the university: staff preparation with high qualifications.

In the traditional learning process the level of competence acquired by students depends on many factors, among which the main ones are:

- Organisation of the learning process at all levels of education (from preparing grids and curricula to conducting specific classes)
- Hardware and software equipment
- Ergonomic conditions
- Teaching staff

The position of each university among others is assessed according to the ranking, taking into account the basic activities of each teacher and university in its entirety, such as teaching, research and education.

ODL may consider it as a new teaching technology; it is as good as it extends the learning possibilities of everyone in every life situation, practically without restrictions; however, the teacher's charisma as one of the important motivation factors is lost.

Open learning combined with distance learning will mean that students must become active, almost equivalent to teachers, participants in the teaching system. This is due to two factors:

(1) Under ODL conditions, student preferences have a great impact on the university's market position.
(2) A lack of direct contact with the teacher requires the student to be aware independently create his/her own cognitive process.

Under the impact of these factors, the management system of the educational organisation should take into account the new position of the student and the appropriate motivation model.

In the conditions of a lack of direct contact between the teacher and students, the scope and manner of performing the aforementioned activities (teaching, research, education) significantly change, as shown in ("e-Quality", 2004). The difference is due to objective reasons: the learning process, where the accumulation of knowledge resources guaranteeing a professional career has a fairly long period of time, the current one is increasingly not working. Examples are such areas as IT, energy, software and banking (Ciszczyk, 2006).

In the traditional learning system, the learner plays a passive role and acts as a carrier of acquired knowledge. Personal characteristics can only be included in direct contact with the teacher. Motivation is done by scoring the evaluation of acquired knowledge (testing, exams, etc.). Didactic materials play a secondary role compared to the teacher, who acts as an intermediary between the source of information and the student's "cognitive" process. As a management object, the student's cognitive process is characterised by high entropy, but through direct contact, through information exchange, the teacher, on the basis of his/her own competence, constantly reduces entropy, that is, he/she manages the cognitive process within certain limits. The effectiveness of management depends to a large extent on the intensity of direct exchanges of information between the teacher and student and the manifestation of the teacher's charismatic power.

In (Kushtina, 2006) it has been shown that the role of didactic materials in ODL conditions is increasing considerably and is derived from the assumption that knowledge is structured information in accordance with the purpose and level of learning.

Processing of information into knowledge takes place by means of its internal structuring – coding and clustering and creating some kind of internal semantic networks, which we can consider as subjective ontologies.

For the didactic material to perform the role of an intermediary between the source of information and the cognitive process of the student in any measure, it should contain the ontology of the subject of study, developed by the teacher. A more detailed construction of learning materials based on the ontological model of the field has been described in (Kushtina, 2006).

The labour-intensive preparation and provision of ODL didactic materials by the relevant IT environment (repository) requires a lot of intelligent and time-consuming efforts by the teacher. The IT aspects of the repository were raised in (Coombs *et al.*, 1970). This implies the need to motivate the teacher to supplement his/her traditional duties with the development and monitoring of the repository.

The second difficulty is the need to motivate the student to quite strongly engage in the process of self-teaching, which will guarantee gaining a level of competence compared to traditional teaching.

Traditional distance testing loses its meaning as an instrument of motivation because it is deprived of all derivatives of direct contact with the teacher (cognitive, emotional, etc.).

A universal way to raise the active position of a learner is a "game", understood as active cooperation, the result of which will be of interest to both the teacher and students (players). In the IT sense, this means that the interests of each of the collaborating participants should be described as student motivation functions that make up one purposeful function.

The repository is the result of this cooperation: From a didactic point of view, it is open to all storage for didactic materials, including ontologies, tasks and examples of their solutions. From a scientific and research point of view, it is a university's knowledge resource in which copyright is reserved. From a software and technical point of view, it is an IT system based on an appropriate network platform.

3.2.2 Stating the Problem of Motivation in the Specific Learning Situation of ODL

We will consider a model of motivation in ODL conditions as the scenario of the game (interaction) of the teacher and students while performing tasks in a specific learning situation aimed at raising the student's involvement in the topic of the task and expanding the repository with new tasks and ways to solve them.

The teaching process in every learning situation includes didactic, research and educational aspects and takes place at the following levels: intelligent, information and IT. At these levels, each teacher and student has his/her own role and degree of involvement. At the intelligent level, assumptions and task-solving are made; at the information level, information is exchanged between participants of the teaching process; and the IT level is characterised by the organisation of the repository and the ability to use it.

The teacher's role is to develop an ontological model corresponding to the topic of the learning situation, show source information, formulate tasks and present methods and examples of their solution in the repository. Tasks are created on the basis of ontology and differ in their degree of complexity.

The student's role is to choose a task and solve it. The final assessment depends on the correct solution and the complexity of the task. The task done by the student and highly rated by the teacher becomes placed in the repository and will serve as an example of performance for other students. All materials placed in the repository have copyright. In this way, the student participates in the didactic activity, and we assume that it will serve to increase his/her self-esteem, which has a positive impact on teaching activity, that is, it will constitute the student's motivation function.

At the same time, filling the repository with a wide spectrum of tasks solved by students serves the teacher's satisfaction as a labour-intensive, effort-intensive, intelligent stage of the initial repository preparation. This will also contribute to the teacher's motivation function.

Work of the teacher and student with a repository can be of a research nature. We assume that the thematic content of the repository coincides with the teacher's research interest, which will result in the appearance of tasks in the repository that differ from typical tasks in their complexity. We can assume that for some students, participating in joint scientific research is a challenge, and the possibility of participating in scientific results even more.

The educational aspect will be understood in such a way that increasing the repository is a joint success of all participants in the learning process, and the copyright notice for both teachers and students displays and visualises the participation of each participant and shows the joint success.

The feeling of synergy motivates one to develop cooperation skills and tolerance. Distance collaboration requires a more logical formulation of questions and answers. All this corresponds to the interests of the teacher and students.

Figure 3.2 presents a diagram of this scenario of completing the task repository in the ODL system.

The learning situation can be characterised by an appropriate ontology of the subject, some of which overlaps with the ontology of the task (in the sense of the number and depth of concepts used).

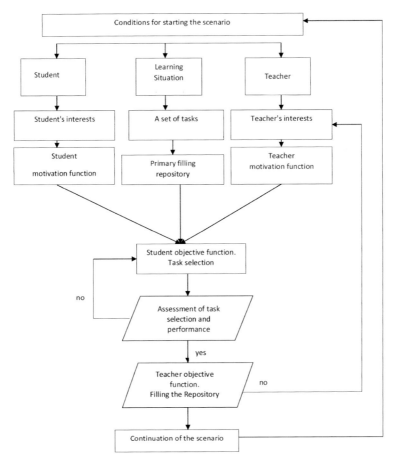

Figure 3.2 Scenario of filling and using the repository.

The teacher's interest is to maximally fill the repository with solved tasks of varying complexity for each educational situation assigned to him/her.

The criterion for placing a task in the repository is determined by the teacher: the degree of task complexity and its relevance to the teacher, graphic quality, language correctness, etc. The possibility of realising the teacher's interests is limited by his/her resources: time in quantitative and calendar terms and other unformalised preferences.

It follows that the motivation model in the ODL system should include parameters describing the activities of each of the interested parties (student, teacher). The measure of the effectiveness of their cooperation will be demonstrated by the increase in knowledge in the repository, which can be assessed by the intensity of its completion with effective task solutions.

When developing the motivation model, an important factor must be taken into account: the stochastic nature of the arrival of students and the stochastic nature of the performance of tasks, which results mainly from the individual learning mode and probabilistic character of student motivation parameters.

The motivation model regulates the student's choice of tasks to perform in a certain subject on the basis of his/her own motivation function, taking into account the teacher's requirements and preferences. The whole process from the moment of formulating the tasks, to the moment of their assessment, placing in the repository and creating new tasks waiting for the next group of students ready to perform them, we describe using the game scenario.

The scenario presented in Fig. 3.2 is universal for every learning situation, aimed at acquiring not only a portion of knowledge, but rather a competence based on it. Game scenario modelling needs to formulate the motivation functions and goal functions of the game participants in relation to completing the repository.

3.3 Team Project Process Oriented on Acquiring Competence

Methods of team creation for the consortium implementing scientific research project involve the use of multiple-criteria decision-making. One of them is the criterion of having competencies. In the analysis of the criterion the team-candidates for the project are compared from point of the view of having competencies required to solve the task. Teams that have all the competencies necessary to execute the project are preferred. If a team does not have all the required competencies, it must incur a cost related to getting the missing knowledge and skills. It can be stated that the usefulness of a team according to the criterion of having competencies is inversely

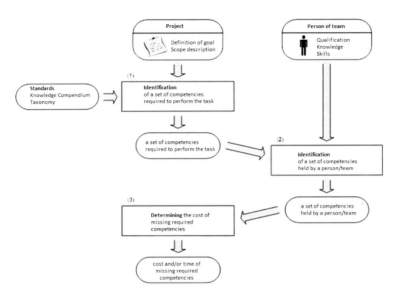

Figure 3.3 Model of the process to analyse the competence to perform a project task.

proportional to the cost of obtaining the missing competencies required for effective implementation of the project (Różewski and Zaikin, 2015).

The model of the process to analyse the competence to perform a project is shown in Fig. 3.3. The whole process to analyse the competence to perform a project task consists of three stages:

1. *Identification of a set of competencies required for the project task*: First, on the basis of the description of the objectives and scope of the project, all the competencies needed to effectively implement it must be identified. In the simplest case, if the project task is one of the typical, frequently realised project tasks, it is possible to find standardised competencies using one of the existing standards or norms. When the project task is atypical and any standards of competence do not exist, the skills necessary to implement this task may be identified through expert analysis. The experts making the analysis may base their own experience and various sources of knowledge in the field of the project task. These can be all kinds of books, articles and compendia of knowledge, whose analysis can help identify

the typical competencies related to the project task domain. For example, the competencies related to solving mathematical problems or solved using mathematical methods can be identified on the basis of the classification of mathematical terms (called mathematics subject classification), which categorises as a taxonomic mathematical discipline (Broens and de Vries, 2003).

2. *Identification of a set of competencies the team has*: On the basis of a set of competencies required to complete the task identified at the first stage, it is then possible to identify these competencies in team-candidates. This can be achieved by analysing their experiences in the form of previously completed projects, experimental research, publications, reports, etc. The most reliable source of knowledge about the competence of a team is to analyse the formal qualifications of its members, or obtained diplomas, degrees, certificates of completion of training, etc.

3. *Determining the cost of obtaining the missing competencies required for the project task*
   : Quantitative analysis of the cost extension of the competencies by comparing the sets of the competencies of the team with the set of skills required to accomplish the task. This analysis is performed using mathematical models of competencies outlined in Chapter 6. The costs of extension of the competencies for each team-candidate are used for comparing them according to the criterion of having competencies in the proposed method of choice of teams for the consortium (Kushtina *et al.*, 2009).

All activities in a specific environment are based on four basic principles that must be implemented in the functioning of the education model:

- Team project work in an open competence-based learning environment
- Ontological approach to the structure of didactic materials
- Motivational modelling and collaborative behaviour in the project process
- Incentive model of the project process in terms of game theory

Let's consider the following problems of the team project process oriented on acquiring competence:

- How should one represent the structure of didactic materials?
- How should one motivate project process participants?
- How should one realise the learning/project process?

Objective reasons arising from the principles of the Bologna Declaration require a high level of openness in the learning system. This leads to the need to consider the aforementioned problems at the level of a general, little formalised description of the system's functioning. In this context, the process of continuous adaptation of the learning system to the current requirements of the labour market, technology and scientific and research development was specified and described at the meta-knowledge level. Each of these problems is considered on the basis of the theoretical and practical principles relevant to it (Fig. 3.4).

The first problem is analysed on the ontological modelling canvas. Ontology as a method of knowledge representation through specification of concepts and relationships between them gives the possibility to conceptualise a given knowledge domain. It is possible to specify concepts describing theoretical, procedural and project knowledge in order to create a hierarchy of those concepts and to show the sequence between them.

The second problem is examined, on the one hand, in terms of cognitive science and, on the other, in terms of motivational modelling. The principles of cooperation of specialists during the development of the project process have been described on the basis of game theory. The interpretation and solution of the developed model can be conducted on the basis of games theory, which allows one to study the activity of a system, depending on the players' behaviour.

The third problem concerns the specifics of using the educational organisation's network resources in performing all the processes that make up the functioning of the ODLS. The problem is analysed on the baiss of the theory of queuing systems. The collaboration process between students and teacher can be interpreted as a queuing system, which has the following goal: there is a possibility

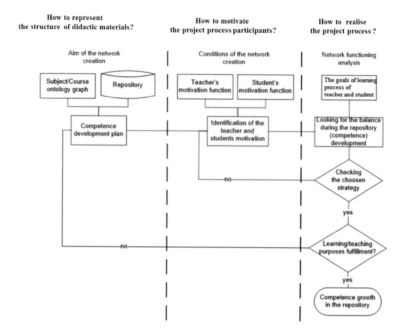

Figure 3.4 Components of the learning process oriented on acquiring competence.

to compare the expected costs that the teacher will bear, considering the assumed repository development (working time) with the results achieved in the didactic process (number of students with a high level of competence participating in the repository, number of students with an average level of competence etc.).

3.4 Models of Competence in the Open Distance Learning System

3.4.1 Types of Competence

Due to the constant development of expert systems (ESs), the expansion of the area of application of decision-making algorithms and the emergence of new organisational structures that have been defined in the world literature as ODL (Zaliwski, 2000), we meet more often the concept of competence in the context of systems.

Competence can appear as a personal property; then we talk about the level of competence of a specialist in a given field (Chen, 2002). But competence can also appear as a group property; then we deal with a class of problems/tasks that need to be solved summary competencies covering several different fields of knowledge (Kushtina and Różewski, 2004).

Two typical task formulations are associated with competence as a personal property. The first concerns the determination and assessment of the required competencies of a specialist at a specific workplace to perform a specific task. The second concerns ODL, where it is necessary to assess the competencies achieved by the student at each stage of independent training. The implementation of these tasks for learning organisations consists in developing human resource management systems. For ODLSs it is necessary to develop specialised didactic materials that correspond to the learning objectives and topics.

We deal with competence as a group property in the following types of tasks. The first type is associated with the effective creation of a group (e.g. project team), operating on the basis of multiple hierarchical games. One of the goals of creating a group may be to find the right relationship between the group's summary competence and the competence required for a given problem/project. In ODL, we meet this type of task when developing a graduate profile by specifying the required summary competencies. The description of the content and structure of competencies needs an analysis of the operation of the decision-making system by a group of specialists using uncertain information and knowledge.

In information systems, the most advanced example of the concept of competence and its broad approach are ESs, where the expert is a single person or a group of specialists (Mulawka, 1997). In ESs, , an expert is the main source of knowledge. Among the most important problems of using expert knowledge, we can distinguish the problem of proper selection of an expert by assessing his/her competence. ESs are used not only to solve a certain task but also to raise the average level of qualifications of specialist staff, which, in turn, requires an assessment of the dynamics of their competence growth.

More seriously, this problem is evident in ODLSs, which is related to the fact that the real criterion for their quality is acquired summary competence in the area of a specific specialty profile. Hence the following problems arising in the implementation of IT systems for teaching students and professional staff: how can we formalise the competency model, what is their measure and when can the competence be used as a criterion for assessing the quality of organisation of the learning process and cooperation of specialists?

3.4.2 Description of the Structure of Summary Competencies

As shown before, an ODLS relies on specialist knowledge in various fields. At each level of management of this system, different sets of summary competencies are needed on the basis of a common knowledge model (Kushtina and Różewski, 2004). In a market situation, an essential ODLS product is educational services, each of which must provide the graduate with the acquisition of competencies required in the labour market. The competitiveness of an educational organisation working in ODL mode depends on how flexibly and quickly it is possible to introduce a change in the curriculum and in the profile of graduates. This problem can be solved in two stages.

At the first stage of strategic management, it is necessary for experts to develop a generalised model of a new profile of specialisation to current market requirements. The input information at the strategic level is taxonomy of theoretical sciences, market requirements for new professions and trends in technology development. As was shown, the basis of competence is knowledge, that is, input information should be transformed into knowledge. This implies the need for experts to transform market requirements into a generalised model of the renewed domain area. The result of such a transformation can be interpreted as summary competencies expressed in the form of:

$$\widetilde{K} = \{\, \overline{D}, \overline{W}, \overline{Z}, \overline{T}\},\qquad\qquad(3.1)$$

where
> $\overline{D}$ are all domain areas related to the required profession,
> $\overline{W}$ is a conceptual model of fundamental knowledge in selected fields,
> $\overline{Z}$ are required skills for the profession sought on the labour market, and
> $\overline{T}$ are existing advanced technologies in selected fields.

At the second stage of planning, it is necessary to find the best processing of the generalised model for the set of curricula containing the thematic arrangement of each subject, divided into theoretical knowledge, laboratory classes and projects, which, in turn, serve as the basis for the development of the required summary competence. Input information at the planning level are a summary competence graph obtained at the first stage, in the form (3.1), and principles and limitations of a substantive and quantitative nature, which reflect the specificity of the university's technical and staff resources.

The result of the second stage is a model of the teaching profile $\widetilde{S}$ in the form of:

$$\widetilde{S} = \{\, P, F, U \},\qquad\qquad\qquad (3.2)$$

where
> P is the subject of teaching $P \in \overline{D}$,
> F is the theoretical knowledge $F \in \overline{W}$, and
> U are skills related to the use of F when used $\overline{T}$.

In the conditions of traditionally understood distance learning, the first stage is the prerogative of the parent organisations and the second stage is carried out at the level of the management of the organisations directly teaching. In the conditions where the educational organisation operates on the market of training services as an independent enterprise, the first and second stages are an internal problem, which is associated with a completely new type of management – *knowledge management*.

Summary competencies are thus a way of presenting information from the environment of an educational organisation, while $\widetilde{S}$ is a description of the programme of introducing a new teaching profile, a conception of responding to changes in the environment. The development of competencies – similar to the development of the profile $\widetilde{S}$ – requires the cooperation of specialists/experts in various fields, but the competitiveness of the university is directly influenced by the quality of developing the internal concept and its implementation. Finding the right mechanism for processing the internal concept into specific subject curricula and developing competency assessment procedures acquired by students during teaching will provide an opportunity to assess the quality of the student's open life cycle.

3.4.3 Organisation of Collaboration of Specialists in Developing the Project

The problem of organising a collaboration of specialists when developing a specialty profile (graduate profile) is similar to the problem of selecting staff for project development, when the project goal, time and financial limitations have already been established. The project's success guarantee consists of:

1) Sufficient summary competencies of specialists involved in the project
2) The way they are organised (specific scenario, game model)
3) Assessment of information certainty that participants will be able to use both outside and inside the project through mutual communication

The development of a formal model that will take into account the listed constituent factors is complicated and needs an explanation of the source that predetermines the scale of specialists' competence or a description of how to describe them. In innovative situations, which include the problem of developing a specialty profile in ODLS conditions (the need to respond to market requirements), it is not possible to rely on a fixed scale of competencies with an orientation on graduate profiles.

For this it is important to find the right method to determine the required competencies. The selection of competent partners, regardless of the criterion for assessing the results of solving this problem, should be considered within the framework of motivational management. Motivational management (as opposed to institutional and information management) consists in creating a stimulation system aimed at achieving maximum competencies at a minimum cost.

Tasks of this type are considered in game theory (Owen, 1975). An analysis of the interests and goals of participants in the process of developing a specialty profile shows that the solution to this problem can be implemented in the form of a cooperative game, where the players' goal is the aggregate profit of a stable coalition (Malawski *et al.*, 2004).

3.4.4 Determining the Requirements for the Team Implementing the Specialty

As mentioned in Chapter 2, the set of competencies is the set of knowledge, information and skills necessary to solve a problem effectively (creating specialties). For a given problem E, *a truly needed competence set* – CN(E) – can be determined, and a set of competencies currently possessed by the *acquired competence set* – CA(E). To solve the problem E, the individual/unit needs to expand their competencies, which always requires a certain time and effort. The set CN(E) is contained in the space of the problem domain D. The space of the problem domain (sc. *habitual domain*) contains all competencies related to this domain.

Let's denote:

E is the decision problem (development of a specialty),

D is the space for all competencies in the field of the problem,

CN(E) = $\{x_i\}$ – a set of competencies necessary to solve problem E, and

CA(E) = $\{x_i\}$ – the currently possessed set of competencies.

The following relationship exists between these sets of competencies:

$$CA(E) \subseteq N(E) \subseteq D. \tag{3.3}$$

For any pair of competencies a and b from space D, if competency b can be achieved from competence a in finite time, there is a relationship for them $a_t \to b$. Due to the existing relationships between competencies, the space of the problem domain can be mapped by a digraph (Yu and Zhang, 1990; Wang and Wang, 1998).

It is assumed that any competence b can be achieved from any competence a if there are no time limits. Therefore, you can define the function t described on the Cartesian product $D \times D$, which assigns a value from the set of real numbers to each pair of competencies and has the following features (Yu and Zhang, 1990):

$$(*) \qquad t(a, b) \geq 0, \; t(a, b) = 0 \text{ if } a = b \qquad \qquad (3.4)$$

$$(**) \quad t(a, c) \leq t(a, b) + t(b, c) \; \forall a, b, c \in D \qquad \qquad (3.5)$$

Similar to the function $t(a, b)$ determining the time cost of achieving competencies, for space D you can also define functions related to any other cost. Most studies discussing the issue of sets of competencies relate to methods of effectively expanding competencies taking into account time and cost constraints.

The extending process is understood as a path $\Gamma = (x_{k_1}, x_{k_2}, ..., x_{k_n})$ that does not contain cycles, stretched on a graph defined by the competence set $CN(E) \setminus CA(E) = (x_1, x_2, ..., x_n)$.

To examine the process of extending competencies, the cost of achieving a new competency from the currently possessed set $CA(E)$ should be determined. Assuming that m is the cost function defined on $D \times D$, you can define a function M that determines the cost of obtaining a new competence from the level of any currently possessed set of competencies (Yu and Zhang, 1990):

$$M(A, x) = min\{m(a, x) | a \in A\}, \; A \subset D, \; x \in D. \qquad (3.6)$$

The optimal extending process $\Gamma = (x_{k_1}, x_{k_2}, ..., x_{k_n})$ is obtained by using the principle of matching such competence $x_{k_i} \in CN(E) \setminus [CA(E) \cup \{x_{k_1}, x_{k_2}, ..., x_{k_{i-1}}\}]$ to the set $CN(E)$, which minimises the cost at a given stage of extending:

$$M(CA(E) \cup \{x_{k_1}, x_{k_2}, ..., x_{k_{i-1}}\}, x_{k_i}) = min \qquad (3.7)$$

In addition, the process of extending sets of competencies is expanded to introduce a creativity factor $\alpha \in (0;1\rangle$ that determines the speed at which the unit of the selected team acquires competencies. The creativity coefficient is taken into account by the following modification of the formula (3.6):

$$M(A,x) = min\{\frac{m(a,x)}{\alpha}|a \in A\}, \ A \subset D, \ x \in D. \tag{3.8}$$

Research projects emphasise that in addition to achieving the project's goal, participants also develop by gaining new competencies. Therefore, the requirements for the team of research units participating in the project should be determined so that part of the competencies required to solve the research problem can be acquired while working on the project. The new competencies acquired in this way will benefit the participants, and their number will be the basis for making the decision on the selection of the group implementing the project. Of course, such a group must have an initial set of competencies that will enable the implementation of the research project, without exceeding time and budget constraints. The set of competencies necessary to participate in the project C_{min} is determined for a given duration, budget and specific value of the creativity factor using the criterion (3.7).

3.4.5 Statement of the Task of Selecting Competent Partners

The process of selecting partners to participate in a research project whose purpose is to solve the problem E in the field D is thus to solve the following task:

For the given:
   $P\{p_i\}$ – set of potential participants,
   $C(p_i)$, $i = 1...n$ – set of competencies for each of the potential participants,
   $\alpha(p_i)$, $i = 1...n$ – competence ratio of each of the potential participants,
   t – time allocated to project implementation,
   b – project budget,

α – minimum value of the creativity factor, and

$C_{min}(E)$ – set of competencies required to participate in the project, based on restrictions time t, budget b and creativity α,

Determine:

P_k – a set of partners with competencies to implement the project E; $P_k \subseteq P$, whose aggregate competencies of all partners include the set $C_{min}(E)$:

$$\bigcup_{P_k} C(p_i) \supseteq C_{min}(E). \tag{3.9}$$

Possession competence means that partner p_i meets the following criteria:

(1) Possessing the right level of creativity

(2) Possessing the competencies required to participate in the project

(3) Impact on the formation of coalitions capable of implementing the project

3.4.6 Analysis of the Partner Evaluation Criteria

The creativity criterion is related to the minimum creativity coefficient α adopted by the institution that selects partners for the project. The value of α is used to determine the set of competencies required to participate in the project $C_{min}(E)$, so even if the group of partners implementing the project has all the competencies specified by $C_{min}(E)$, it must have individual creativity factors $\alpha(p_i)$ not less than α. Otherwise there will be a risk of exceeding the time or budget limit of the project:

$$\alpha(p_i) \geq \alpha, \quad \forall p_i \in P. \tag{3.10}$$

Each potential participant in the project for which there is no dependence (3.10) is considered incompetent and should be removed from the set P.

The criterion of having the competencies required to participate in the project is to check potential participants for their competencies from the set of competencies required to participate in the project $C_{min}(E)$:

$$C(p_i) \cap C_{min}(E) \neq \emptyset. \tag{3.11}$$

A potential participant in the project who does not have any competencies from the set $C_{min}(E)$ presents no value from the point of view of condition (3.9) and should be removed from the set P.

The criterion of influence on the formation of coalitions capable of implementing the project. Despite meeting the first two criteria, some potential participants do not have any unique competencies, that is, in each possible case of partner selection their competencies may be replaced by the competencies of another project participant. Detection of this type of partners is possible, thanks to the solution of the cooperative game in the form of a function characteristic of the set of partners P. The characteristic function of such a game should be defined as follows:

$$\nu(S) = \begin{cases} 1, & \text{if } C_{min}(E) \subseteq C(S), \\ 0, & \text{if } C_{min}(E) \not\subseteq C(S) \end{cases} \tag{3.12}$$

where

S is the coalition of players (any subset of the set P) and
$C(S) = \bigcup_S C(p_i)$ is a set of competencies of coalition S.

A solution to this formulated game can be found by calculating Shapley's value (Malawski *et al.*, 2004; Straffin, 2004; Kałuski, 2002; Owen, 1975). The algorithm of calculating Shapley's values checks all possible orders of joining partners to the project team. For each potential participant in the project, it is checked individually how much has increased – for a given order of joining – the value of the coalition after joining the participant.

This value is 1 when the joining participant finds a coalition that does not cover its total competencies $C(S)$ of the set $C_{min}(E)$, and adding its individual competencies $C(p_i)$ to the coalition competence set $C(S)$ will fully cover the set $C_{min}(E)$. However, this value is 0 in all other cases, when the joining participant finds a coalition that already covers its own competence set $C_{min}(E)$ or when it finds a coalition that does not cover the set $C_{min}(E)$ and after its attachment there is still no coverage.

The result of the game calculated in this way will be imputation:

$$X = (x_1, x_2, ..., x_n), \quad n = |P| \tag{3.13}$$

whose elements x_i determine the frequency of 1 for a given player. The higher the frequency, the greater the partner's contribution to the formation of groups of partners capable of implementing the project. However, when a given participant's frequency is 0, it means that it has no impact on the formation of groups of partners capable of implementing the project and does not have any unique competencies supplementing the competencies of other project participants. Such a participant should be removed from the set P. We assume that the higher the level of summary competence of the team selected during the game, the higher the quality level of the specialty profiles developed by them.

3.4.7 The Problem of the Quality of Developing and Implementing the Teaching–Learning Process

The problem of developing the concept of a quality management sub-system within the information system supporting the management of the enterprise consists mainly in the analysis of existing quality control methodologies. A comparison of this type of methodology will be considered in terms of stability of the following two factors:

1) Production process
2) Organisational structure

The result of this comparison is shown in Table 3.2.

One of the main areas of research in the field of ODL is the issue of quality (McGorry, 2003). This problem is also raised by the e-Quality project ("e-Quality", 2004) financed by the European Union, which carries out the mission of developing quality standards for emerging open educational organisations. By shaping the profile of scientific programs, the European Union influences the concentration of European research areas on selected issues.

Table 3.2 Comparison of quality control methodologies

Production process	Organisational structure	Quality control method
Stabilised	Stabilised	Approach based on ISO methods, appropriate for material production in industrial enterprises.
Innovative	Unstable	Process approach consisting in the identification of processes occurring in the organisation by defining and describing the inputs and outputs of given processes. Approach characteristic of the innovative situation of creating a project in a multi-project environment, for example creating a new specialty in an educational organisation.
Poorly formalised	Stabilised	Approach based on reporting. Due to difficulties in describing processes, quality is controlled by creating accurate descriptions of process behaviour in the form of reports. It is important that the report have a structure that allows it to accurately express the existing situation. We find this type of situation in organisations with a social profile, for example foundations.
Poorly formalised	Unstable	On the basis of the concept of games and scenarios, we describe processes that take place in an open environment. Because, due to the open environment, we are not able to predict the scope and depth of future tasks, we base quality control on the analysis of the competencies of people performing a given task. We examine employee competencies at the individual and summary levels, where the overall picture of cooperation that allows the organisation's goals to be checked. An example of the described situation is an organisation operating in ODL conditions.

The most important goal of the programme is to develop standards for creating didactic materials and for managing the distance learning environment, while maintaining quality criteria. Let's look at the genesis of the issue of quality in the European distance learning space.

According to the findings of the Bologna Process (June 1999), quality was one of the basic mechanisms involved in creating the European Higher Education Area by 2010. This direction was confirmed during meetings in Prague (May 10, 2001) and later in Berlin (September 19, 2003), where the quality issues were specified by systematising the foundations of the discussed mechanism. The quality assurance mechanism ("European Association for Quality Assurance in Higher Education", 2016) is based on the assessment carried out in the following modes: evaluation, accreditation, audit and benchmarking.

Individual elements aiming to improve quality operate according to a four-stage model. The first level assumes quality control independently by each institution by shaping its own procedures and standards aimed at maintaining and improving quality. The next step is the self-assessment process, which takes place in set periods and is based on a prepared questionnaire (survey). The collected documents are the basis for the next level, assuming the visit of external experts (in Poland they are members of the State Accreditation Commission). A report based on the collected data is created. Publication of the report is the last stage of the model. In further considerations, we will move away from quality analysis from the organisation's point of view to process approach, focusing on the process of learning–teaching.

The issue of knowledge transfer can be discussed in the context of intangible production processes. According to (Korytkowski and Zaikin, 2004), an intangible product (e.g. a distance learning course) is a product with a digital form that is the result of collaboration between specialists. Learning in ODLS conditions consists of a number of intangible products (including the specialty profile), which are the result of intangible production. An important problem of quality analysis in intangible production is the difficulty in quantitative assessment of their properties such as depth and topicality of knowledge.

Table 3.3 Main processes in distance learning with their assigned roles (Zaikin and Różewski, 2005)

Processes	Sample roles
Planning the teaching process	Manager of the rules of the teaching process Teaching Content Creator
Administration (management) teaching process	Administrator of the teaching process Coordinator Technical Administrator Consultant
Assessment and evaluation of learning outcomes	Students evaluator Developer
Production of didactic materials	Designer of teaching materials Contractor of teaching materials Audio and video specialist
Support of the teaching process	Pedagogical support Technological support Tutor

The process approach allows the analysis of the quality issue from the point of view of quality management, which takes place on the basis of separate processes. As a result of the analysis of the ODL phenomenon, the following processes were separated as part of the e-Quality project: (1) planning, (2) administration, (3) evaluation, (4) production of didactic materials and (5) student support.

Each of these processes has been assigned roles that characterise the actions that must be performed to achieve the intended goal. The role status is more flexible to situations in which the interaction scenario is strictly defined, which is possible in a stable situation. When dealing with an innovative situation, a process approach based on establishing the relationship between the process and the performance of specific roles does not seem to be adequate. In addition, defining roles at the implementation stage of the open teaching system is difficult due to different educational systems and cultural differences. The prepared set of roles (e.g. Table 3.3) is a result of compromises and in most education systems requires adaptation to existing conditions. Each of these roles can be described using the standard Rational Unified Process (RUP). Originally, RUP is used to plan and manage projects related to the construction and implementation of various types of software (Henderson-Sellers *et al.*, 2001).

In the e-Quality project, RUP was used to describe roles using the following structure:

⟨ *role, activity, artifact, additional elements* ⟩

- *Role*: based on the ODL idea and the local education system.
- *Activity*: a set of actions performed by a given role, which are motivated by a common goal (e.g. planning the structure of teaching material).
- *Artifacts*: man-made things; a particularly important artifact is information about how it is created, modified and used by individual processes.
- *Additional elements*: auxiliaries that do not directly participate in the process, e.g. guides.

After presenting the process approach used in the e-Quality project to analyse the ODL idea, let's take a look at exactly two selected processes: the design process of teaching materials and the student support process. These are processes that follow the already developed specialty profile.

The analysis carried out by the e-Quality consortium showed that the processes discussed are key when it comes to the quality of the entire teaching process within a separate educational organisation. Finding common views on the issue of the quality of these processes will serve as a starting point for developing quality standards for the entire *teaching and learning process*. In addition, investments aimed at improving quality in the processes discussed are relatively small.

The design process of didactic materials is aimed at preparing materials (resources) for learning a given subject. In distance learning, the process of preparing didactic materials takes on special significance due to a lack of direct contact between student and teacher. The student, thanks to the didactic materials, discovers knowledge by himself/herself, striving to build appropriate (correct) thought structures in his/her mind. The design process of didactic materials is product-oriented and usually intangible. Each created material is unique, prepared to meet the assumed requirements of a given course, but it will still be able to contain learning objects modules previously developed for other courses and placed in a

Table 3.4 Comparison of student support processes and preparation of didactic materials (Zaikin and Różewski, 2005)

Process features	Design process of didactic materials	Student suport
Intended use	Product orientation	Customer orientation
Production type	Unit production	Mass production
Character	Deterministic	Random
Modeling methods	Gantt graphs	Simulation
Criterion type	Non-formal criterion	Quantitative criterion

repository. Especially, this type of task is handled in a set time schedule. To analyse the quality of the didactic materials preparation process, we will base it on qualitative features such as usability (ergonomic aspect), competence (information aspect) and structure (cognitive aspect).

Another look at the idea of quality is presented in the student support process, which covers activities such as assistance offered to the student to solve technical problems or pedagogical support. The support is based on the personal infrastructure and the student's software and equipment. Examples of infrastructure are telephone technical support systems (*help-desk*) or a digital library of teaching materials.

The discussed process is focused on a student (as client) who reports to a given service at a random moment of time and takes it for a random period. In this case, the key criterion is the effectiveness of the environment, which can be calculated by analysing the appropriate times and costs (e.g. the total time when the student requests access to resources). The calculation of these types of quality indicators needs an optimisation model for a closed network consisting of servers (educational services): teacher, course, administration and students (Zaikin *et al.*, 2000). As you can see, the process approach is more transparent on the one hand and can be based on the experience of previous practices and/or use of computational methods. On the other hand, in this situation we will not be able to assess the summary competencies used in the design of these processes. The need to consider the problem in terms of summary competence can be seen in Table 3.4.

The final learning outcome – competencies acquired through the student – depends on the quality of both two processes, and additionally, the quality of student support depends on the quality of didactic materials (e.g. time and number of consultations and trainings). The table has clearly shown that the analysis, modelling and interpretation of the quality assessment method for these processes belong to different abstraction zones, which means that the final result of their implementation must be assessed in terms of total competence.

The conception of a system approach to quality analysis in ODL processes shows the difficulties faced by the analyst. Quality is achieved by ensuring high accuracy of calculations, which due to existing cultural and system differences in individual education systems requires a lot of work. In addition, looking generally at the ODL process, we see the different nature of its subprocesses – each of them has a different purpose and criteria for success. In the hierarchical scheme of the ODL system they belong to different branches of the graph (Kushtina and Różewski, 2004).

Generally speaking, the following models and their mutual compatibility constitute the criterion of the quality of the teaching–learning process at the student's open life cycle level:

- Knowledge representation model
- Competence acquisition model
- Knowledge management model

The main criterion of the ODL system operation can be interpreted in the context of ensuring the quality of operation of separate subprocesses that maximise the fulfillment of the individual demand for the student's time and teaching mode and minimise differences with the traditional teaching environment as well maximise the possibility of obtaining a certificate of learning results.

Chapter 4

Ontology Modelling in Open Distance Learning

4.1 Ontology as a Method of Knowledge Representation

Ontology is a tool for describing the field of knowledge, which provides the basis for modelling the content of concepts and relations between them.

Ontology should enable the class of objects necessary for a complete description of a specific field of knowledge to be identified and to draw conclusions about the processes taking place is to be a tool for describing the field of knowledge that gives the basis for modelling the content of concepts and relations between them.

Heylighen developed this approach by recognising ontology as an effect of an abstract analysis of the concept of a conceptual system that can be presented in the form of a graph whose nodes map the basic concepts, while arcs correspond to the relations between concepts (Heylighen, 1990).

The result of describing a specific field, according to the accepted reasoning method is an ontological model.

Open Distance Learning: Fundamentals, Developments, and Modelling
Oleg Zaikin
Copyright © 2023 Jenny Stanford Publishing Pte. Ltd.
ISBN 978-981-4877-55-8 (Hardcover), 978-1-003-13261-5 (eBook)
www.jennystanford.com

Definitions of Ontology

- Ontology is a formal, unambiguous specification of shared (common) conceptualisation (Studer et al.,1998)
- Ontology defines the basic terms and relationships that form the dictionary of a given subject area, as well as the rules for combining the terms, as well as expanding the dictionary (Neches et al., 1991).
- The 'light' ontology can be compared to taxonomy,
- The 'heavy' ontology models the whole domain, defining more limitations to domain semantics.

In Table 4.1 the history of the science of ontology and the contribution of different authors in its development are represented.

A concept in ontology is considered as a semantic unit of the domain area. The general concept categories are subject, process, event, state, etc. The structure of a concept can be expressed by its intensional and extensional features. The *intensional* features of a concept represents the content of the concept, i.e. a set of significant features within the domain area, while the *extensional* ones describes the concept volume, i.e. the number of copies/objects belonging to the class (within the boundaries of the problem, tasks). A set of values of significant features within the domain area we define as *depth of the concept.*

Therefore intensional, extensional and depth features of a concept are their basic characteristics and the concepts in a specific domain area are defined as follows:

- They describe the properties of abstract entities.
- They do not change as a result of operational activities.
- They are connected semantically.
- This connection is a loose network.

Current applications of ontologies are most used in such domain areas as knowledge engineering, artificial intelligence, informatics and education. On the basis of an ontological approach, problems such as the following are solved:

Table 4.1 Ontology as a 'philosophy of being'

Author	Date	Contribution
Parmenides	V and IV c. BC	Precursor of Ontology, he recognised the independence of the essence of being from our senses
Aristotle	IV c. BC	The author of *Metaphysics*, which can be called Ontology established a *system of categories* allowing for the classification of everything that can be said about the world
William of Ockham	XIV c.	He described the concept of *'universal'* as a symbol defining all objects of a given type, not a specific copy
Immanuel Kant	XVIII c.	He developed categorisation: *quantity* (unity, multiplicity, generality), *quality* (reality, negation, limitation), *relation* (inheritance, result, commonality) and *modality* (possibility, existence, necessity)
José Ortega y Gasset	XIX-XX c.	He recognised the subjectivity of *perception*
Edmund Husserl	XX c.	He gave rise to a branch of ontological research called *Formal Ontology*

- Knowledge management
- Natural language (NL) processing
- Acquiring information
- Design and integration of databases
- Semantic Web
- Cooperation and communication of users

Examples of projects with a developed ontological approach are:

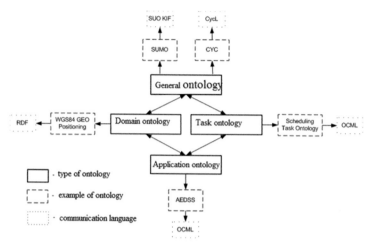

Figure 4.1 Types of ontology.

- TONES (Thinking ONtologiES), OntoGov (6th Framework Program)
- FUSION project Knowledge Engineering, (7th Framework Program).

4.1.1 Classification of Ontologies

The first ontologies built for the needs of knowledge engineering began to emerge in the 1980s. The intensive work on the ontology resulted in the development of its various categories. The division proposed by (Guarino, 1998) includes:

- General (top-level) ontology
- Domain ontology
- Task ontology
- Application ontology

These types of ontology are represented in Fig. 4.1. Each of them has a different level of generalisation:

- *The domain ontologies* describe knowledge characteristic of a specific field, e.g. medicine, pharmacy and music law. The terms used to describe them are the result of the specialisation of concepts defined in general ontology.

- *The task ontology* describes a dictionary related to a specific task or activity, e.g. diagnosis, scheduling, specialisation of terms derived from general ontologies. In contrast to domain ontology, in this case concepts from various fields can be used to solve the problem.

- *The application ontology* contains the definitions that are required for the description of knowledge for individual applications. It expresses and specialises the domain and task ontology dictionary for a given application.

The result of the categorisation of ontologies is two different approaches to the development of ontologies, derived from the philosophical sciences:

(a) From the general to the detail (top-down approach)
(b) From the detail to the general (bottom-up approach)

The first approach pre-supposes the development of ontologies by first finding the general classification of entities common to all fields. However, there is a low probability of finding a satisfactory general ontology that would meet the assumptions of all application contexts and domains, but on the other hand, such a solution would not only give the right shape to the general ontology but also allow for inference between specific domains.

Applying in accordance with the sequence *from detail to the general* proposes to first develop domain ontologies and only on their basis draw conclusions about a more general ontology. The lack of a unified and universally accepted method of creating ontologies for specific fields means that there is a risk of the inability to build general ontologies by generalising concepts. Nevertheless, each of the categories of ontology distinguished has already developed and functioning reflection in reality, formalised with the use of ontology description languages, mentioned on Fig. 4.1.

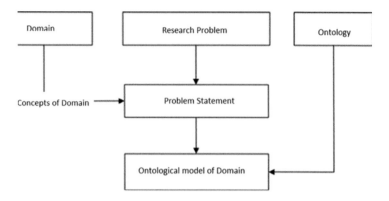

Figure 4.2 Ontology and the ontological model.

4.1.2 Generalisation of Ontology: Example

The introduced categorisation allows for appropriate mapping through the ontology of the way of thinking, and then the formulation of the problem and its solution. It is necessary to take into account the fact that not only the field of problem origin is considered, but also the field from which the problem is to be solved. When using ontologies to solve a certain problem, it is necessary to go through successive levels of its generalisation in order to be able to interpret a specific field of knowledge in terms accepted for this field or compatible with the chosen tool to solve the problem (*mapping of concepts*). An example of ontology generalisation is represented in Fig. 4.2.

Let us consider this way of proceeding on the example of the term 'process'. The first level of generalisation is a general characteristic of the problem and allows its formulation. If we are moving in the sphere of process modelling, then the next categories of ontologies have to determine the field that these processes concern, for example, domain – *production systems*. Due to the fact that we can deal with different types of processes occurring as a research object, the content of the task is determined, for example, *stochastic processes*. Only the whole gives the opportunity to find an approach to solving the problem – *simulation modelling*.

This solution needs analysis because you need to find a way to solve the problem. One should go to the second degree of

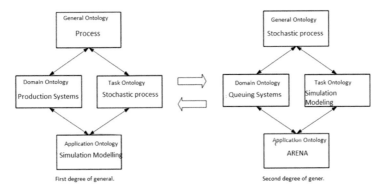

Figure 4.3 Example of generalisation of ontology for the concept 'process'.

generalisation of ontology, in which stochastic processes become the main category of considerations. *Stochastic processes* can be described, for example, in the field of queuing systems, and as such become a further object of research. It is necessary to place a task, which is a starting point for the transition to the next level of categorisation. Defining the simulation modelling task as an ontology concerns the construction of a specific simulation model. Understanding what type of simulation is to be carried out leads us to a solution to the problem that the *Arena simulation package* might be in this case. An example of ontology generalisation for the concept 'process' is represented in Fig. 4.3.

The levels of generalisation of ontology for concept 'learning process' are represented in Fig. 4.4.

4.1.3 Methods of Creation and Ontology Languages

In e-dictionaries, data is saved in the database as a single file with the appropriate structure. Each database entry is called a record, while records are made up of fields.

A database index is a data structure that improves the speed of data retrieval operations on a database table at the cost of additional writes and storage space to maintain the index data structure. Indexes are used to quickly locate data without having to search every row in a database table every time a database table is accessed. Indexes can be created using one or more columns of a database

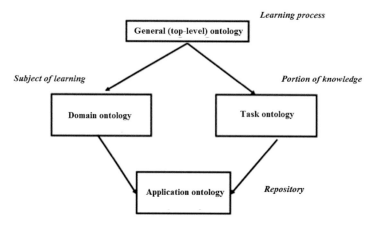

Figure 4.4 Levels of generalisation of ontology for the concept 'learning process'.

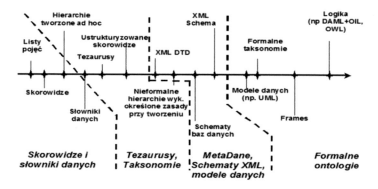

Figure 4.5 Methods and tools to create ontologies.

table, providing the basis for both rapid random lookups and efficient access of ordered records.

A thesaurus (plural thesauri or thesauruses) or synonym dictionary is a reference work for finding synonyms and sometimes antonyms of words. They are often used by writers to help find the best word to an express an idea. Synonym dictionaries have a long history. The word 'thesaurus' was used in 1852 by Peter Mark Roget for his *Roget's Thesaurus*, which groups words in a hierarchical taxonomy of concepts, but others are organised alphabetically or in some other way.

A data model is an abstract model that organises elements of data and standardises how they relate to one another and to the properties of real-world entities. For instance, a data model may specify that the data element representing a car be composed of a number of other elements, which, in turn, represent the colour and size of the car and define its owner.

XML schema is a description of a type of XML document, typically expressed in terms of constraints on the structure and content of documents of that type, above and beyond the basic syntactical constraints imposed by XML itself.

The Web Ontology Language (OWL) is a family of knowledge representation languages for authoring ontologies. Ontologies are a formal way to describe taxonomies and classification networks, essentially defining the structure of knowledge for various domains: the nouns representing classes of objects and the verbs representing relations between the objects. Ontologies resemble class hierarchies in object-oriented programming but there are several critical differences.

A data dictionary, or metadata repository, as defined in the *IBM Dictionary of Computing*, is a "centralised repository of information about data such as meaning, relationships to other data, origin, usage, and format". Oracle defines it as a collection of tables with metadata. The term can have one of several closely related meanings pertaining to databases and a database management system (DBMS): a document describing a database or collection of databases.

In philosophy, the term 'formal ontology' is used to refer to an ontology defined by axioms in a formal language with the goal to provide an unbiased (domain- and application-independent) view on reality, which can help the modeller of domain- or application-specific ontologies (information science) to avoid possibly erroneous ontological assumptions encountered in modelling large-scale ontologies.

4.1.4 Kinds of Ontologies: Examples

General ontologies: They are top-level, generic or common-sense ontologies. These ontologies consist of general knowledge about the world and basic concepts: time, space, state, event, etc.

Example: Ontology Suggested Upper Merged Ontology (SUMO), proposed by the IEEE. It contains various definitions of general terms that can be used to create specific domain ontologies. This ontology integrates concepts found in various ontologies, such as:

- Ontolingua (Fikes *et al.*, 1997)
- Ontology of John Sowa (Sowa and Zachman, 1992)
- Ontologies of Russell and Norvig (Russel and Norvig, 2011)
- James Allen's temporal axioms (Allen, 1983)
- The formal theory of the holes of Casati and Varzi'ni (Casati and Varzi, 1999)
- Barry Smith's ontology of border areas (Smith, 1976),

The formalism adopted for recording this ontology is KIF. The Standard Upper Ontology (SUO) structural level can be used as a logical framework for manipulating collections of ontologies in the object level or other middle level or domain ontologies. From the information flow perspective, the SUO structural level resolves into several metalevel ontologies. In (Kent, 2011) discusses a KIF formalisation for one of those metalevel categories, the Category Theory Ontology.

4.1.5 Criteria for Ontology Creation

- Transparency: The meaning of terms must be comprehensible, objective and complete definitions, if possible, in NL.
- Minimum coding deviation: Conceptualisation must be determined at the level of knowledge, regardless of coding at the symbolic level.
- Extensibility: It must be possible to define new terms based on existing ones without the need to revise existing definitions.
- Coherence: Conclusions can be drawn; conclusions cannot contradict existing definitions.
- Minimal ontological contribution: Defining only the terms necessary to present knowledge consistent with the weakest theory used.

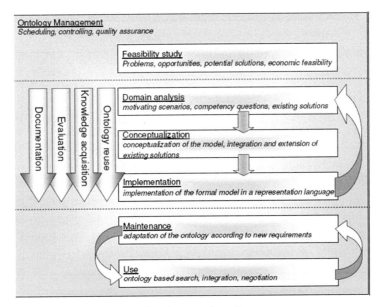

Figure 4.6 Ontology engineering process.

Additional criteria:

- Representation of disjoint and comprehensive knowledge
- Minimal syntactic difference between concepts from one family
- Standardisation of names

Ontology engineering in computer science, information science and systems engineering is a field which studies the methods and methodologies for building ontologies: formal representations of a set of concepts within a domain and the relationships between those concepts. A large-scale representation of abstract concepts such as actions, time, physical objects and beliefs would be an example of ontological engineering. Ontology engineering is one of the areas of applied ontology, and can be seen as an application of philosophical ontology. Core ideas and objectives of ontology engineering are also central in conceptual modelling. Figure 4.6 represents the *ontology engineering process*.

There are a number of basic languages for a description of ontology. Examples of these languages are as follows:

- Traditional languages: Ontolingua, KIF, LOOM, OKBC, OCML, FLogic
- Languages of ontology marking: SHOE, XOL, RDF (S), OIL, DAML + OIL, OWL

As an example, let's consider the *Resource Description Framework* (RDF) language. An RDF data model consists of ordered triples, so-called entities and resources. An entity can be referenced by the Uniform Resource Locator (URL) or Uniform Resource Identifier (URI). *Resources* are elements described in RDF sentences. Binary relations between resources and/or values atomic ones are defined as basic types of XML data. A specific value of the judgement for a given entity is a *complement*. Here is presented an example in simplified RDF notation:

- Author (http://atos.wmid.amu.edu.pl/ d124124) = X
- Name (X) = Michał
- Email (X) = d124124@atos.wmid.amu.edu.pl
- X - Michael's current or virtual URL.

This information saved means that the author of the website with the address given in the task is the X resource, his name is Michał and the email address is *d124124@atos.wmid.amu.edu.pl.*

Figure. 4.7 shows an example of the description of authors as a simplified RDF graph.

4.1.6 Conclusion

- Ontologies reproduce new forms of being.
- IT technologies allow to integrate various ontologies of the field into one new whole.
- The degree of Internet development needs new instruments to integrate thematic resources.
- New sciences and specialties are created: teleinformatics, telemedicine, social networks.

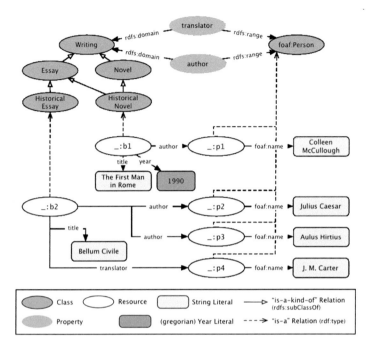

Figure 4.7 Example of a simplified RDF graph.

4.2 Knowledge Representation in Open Distance Learning

4.2.1 Knowledge in the Traditional and Distance Learning Process

The NL is the basic mode of *knowledge representation* and transfer in conventional teaching. However, various types of symbolic/artificial languages can also be used (e.g. Kent (2000)), which gives additional possibilities to enhance the semantic expressiveness of the NL. From the point of view of ergonomics and psychology, the computer environment sets bounds on the NL usage (Chalmers, 2003) On the other hand, the computer environment gives unlimited possibilities for using symbolic and artificial languages and media.

The question is appears: how to achieve, in the open distance learning (ODL) environment, the level of NL accuracy in the knowledge manipulation. The solution to the problem is developing a knowledge model of representation and delivery. Knowledge, as an object and purpose of study, can be divided into two basic kinds: fundamental (theoretical) and operational (procedural). Each kind of training activity, irrespective of the subject domain, demands assimilation of both theoretical knowledge and practical skills.

Fundamental knowledge reflects conceptual thinking and contains new paradigms, problem statements, principles of behaviour, etc. *Procedural knowledge* is necessary for the development and realisation of scenarios, algorithms and the performance of various kinds of operations. Analysing various problems and situations, examined in the learning process, demands simultaneous use of both kinds of knowledge but in different proportions, depending on the complexity of the problem that is being solved. New methods of procedural knowledge representation and ODL technologies allow creating effective systems of computer training (simulators, expert systems, etc.). Such systems set a relationship between practical skills (speed, accuracy, conformity to standards) and the available theoretical base in real time. Therefore, conceptual thinking is shown as follows:

- Abilities of abstraction and generalisation
- Setting associative links
- Inference of a new knowledge on the basis of the available one
- Formulation of paradoxical knowledge which exceeds the existing paradigm

There are no objective quantitative methods of evaluating the conceptual knowledge acquired during training. Therefore, assessment of the quality of learning the conceptual knowledge is based on the objective properties of the domain and techniques of studying it. With the same amount of knowledge in the subject domain and other equal conditions the various techniques of training can affect the development of conceptual thinking in a different degree. The responsibility for the choice of the training technique lies in the hands of the teacher. Therefore, in the conditions of ODL, it is necessary to use the methods of theoretical knowledge representation that give opportunities to present objective knowledge in the subject's domain as well as techniques of studying it.

The analysis of existing models of knowledge representation shows that none of them satisfy the demands of ODL. It is shown in (Quillian, 1968) that the structure of knowledge, which differs from the rules systems, can be presented on the basis of semantic networks. Semantic networks can represent both abstract categories and certain objects. The big obstacle for using a semantic network is the difficulty with formal representation of the semantic relation. Semantic networks are the starting point for such knowledge representation models as mind maps(Buzan and Buzan, 1993), conceptual maps (Mineau *et al.*, 2000; Sowa and Zachman, 1992) and topic maps (Abramowicz *et al.*, 2002; "ISO/IEC", 2000). The models are oriented towards a specific kind of knowledge and a concrete user. Each of them is dedicated to a specific kind of task and a specific domain. The level of universality in the ODL knowledge model should be high. The appropriate model must at least work with one kind of knowledge (e.g. theoretical) in all its contexts. The final knowledge representation system should merge the knowledge manipulation language with the corresponding pedagogical approach, which is used to learn about a subject.

The mind stores knowledge in the form of a concept (Anderson, 2000; Maruszewski, 2002). The concept is considered a structure of mind representation which consists of the proper description of a single-meaning class. The methods of conceptual knowledge teaching take advantage of the concept's communication level between the students. Teacher can use metaphors or other pedagogical methods (Conole et al., 2004). The most important issue is establishing the cohesion state, which is achieved when both, teacher's and student's, minds refer to the same concepts' prototypes simultaneously.

The proposed knowledge model plays the role of a collaboration tool between the teacher and the subject matter expert. In the phases of the learning objectives' creation and the student's knowledge discovery, the following activities are performed based on the knowledge model:

- Learning object's creation
- Semantic border's localisation
- Synonyms chain's creation

These operations give the possibility to edit the concept's semantic from the expert's side as well as the teacher's side. The knowledge model is the basis for the modular didactic materials' formation. The model of each concept is formulated as a logical unit with reference to the media metaphor and other concepts that are stored in the repository. Such approach is compatible with the requirements of the *learning management systems* (LMS) and *learning content management systems* (LCMS), , which are widely used in ODL (Helic *et al.*, 2004). The LMS is a strategic solution designed for planning, delivering and managing all the training events in the company, considering virtual classes as well as the ones taught by an instructor (Greenberg, 2002). The LCMS (Brennan *et al.*, 2001) is what we call a system used for creating, storing and making available (sending) personal educational content in the form of a learning object.

The structure and construction of a learning object are covered in detail by the SCORM standard ("Advanced Distributed Learning Initiative", 2001). However, there is a lack of information or knowledge about the content of the learning object in the SCORM standard. The international research society has been investigating the problem of the learning object for several years (Downes, 2001; Hamel and Ryan-Jones, 2001; Kassanke *et al.*, 2001; Polsani, 2003); however, a general solution has not yet been found. Until now, a set of guidelines and rules has been published (Hamel and Ryan-Jones, 2002). The learning object problem is also discussed in (Lin *et al.*, 2003; Sheremetov and Arenas, 2002; Wu, 2004) and European Union programmes (CANDLE, 2003; PROLEARN, 2004). Authors proposed the described approach as the next stage of the universal distance learning system creation.

4.2.2 Content, Volume and Depth of Knowledge

Reality is defined by an unlimited and diverse set of information and stimulus which attract the human perception system. The cognitive science assumes the natural mind's ability to conceptualise (Gómez *et al.*, 2000). The conceptual scheme exists in every domain (Eden, 2004) and informs about the domain's boundaries and paradigm. On the basis of the ontological approach, one can create a conceptual model of the knowledge domain, where the concepts are the atomic, elementary semantic structures (Benjamins *et al.*, 1999; Guarino, 1997; Studer *et al.*, 1998; Sugumaran and Storey, 2002). From a practical point of view, ontology is a set of concepts $c_1, ..., c_i, ..., c_n$ from a specific domain. The modelling approach, considered as a cognition tool, assumes some level of the examined object's simplification.

In our research the following definition of a concept will be used (Kushtina and Różewski, 2003, 2004): *A concept is a nomination of classes of objects, phenomena, and an abstract category; for each of them, common features are specified in such a way that there is no difficulty in distinguishing every class.* Given a concept's definition makes possible the modelling of the knowledge model for any domain on the basis of the set of basic concepts which were specified by an expert in a verbal way.

The concept ϕ is defined as a tuple:

$$\phi = \langle X, T \rangle, \tag{4.1}$$

where

$N(\phi)$ is the name of the concept ϕ.

X is the set of information, which provided the concept's description. The description is made based on one of the metadata standards (e.g. DublinCore, IEEE LOM).

T is the matrix of the concept's depth. $T = [\hat{t}_{ij}]$ has the following meaning:

$$T[\hat{t}_{ij}] = \begin{cases} N(\phi), & \text{when} \quad i = 1 \wedge j = 1 \\ \text{Attribute } i, & \text{when } i \neq j \wedge j = 1 \\ \text{Object } j, & \text{when } i = 1 \wedge j \neq 1 \\ t_{ij}, & \text{otherwise.} \end{cases} \tag{4.2}$$

All elements of the matrix T belong to the specified domain; as a result, they are included in a single ontology $\{N(\phi), \text{Object } i, \text{Attribute} j, \text{ } t_{ij}\} \in \Omega$, for $i = 1, ..., I$, $j = 1, ..., J$.

Such a concept's definition allows a formal representation of its internal structure and makes it possible to shift from a verbal form to a formal description in the form of an abstraction, which can be organised as a matrix or another graphic structure (Gordon, 2000;

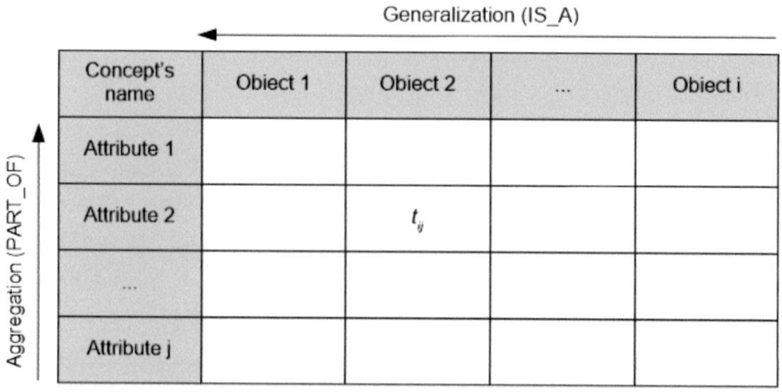

Generalization (IS_A)

Concept's name	Obiect 1	Obiect 2	...	Obiect i
Attribute 1				
Attribute 2		t_{ij}		
...				
Attribute j				

Aggregation (PART_OF)

Figure 4.8 Concept in the form of an abstraction.

Kent, 2000). Therefore a concept can be examined as an abstraction (Goldstein and Storey, 1999), what helps one understand a complex object by decomposing it into uncomplicated parts.

In the case of the matrix (Fig. 4.8), the number of rows equals the number of attributes of the concept and the number of columns symbolises the number of objects merged into one concept's class (Tsichritzis and Lochovsky, 1982). The number of elements in the matrix t_{ij} is called the concept's depth in the specified domain. The coloured field (Fig. 4.8) describes the context of the concept. The basic relations, which occurred between concepts and are defined by the subject matter expert, are the following: homonym, synonym, meaning extension and meaning deepening.

The development of the concept matrix structure is based on the selection and use of existing definitions of concepts. It is known that there are different approaches to creating definitions of the concept, two of which are the most significant in shaping systemic thinking:

1) *Intensional*: revealing the essential features of the term, provided that the species of the defined object is taken into account. ('The tree is a kind of plant with a specific shape of the trunk, crown, roots', 'mammals are an animal species' and 'a computer network is a combination of several computers according to one of the typical topologies using one of the specific protocols for data transfer'),

2) *Extensional*: revealing objects subject to the concept in a specific field ('oak, pine, beech are trees', 'wolf, deer, zebra are mammals', 'LAN, MAN, WAN are the architectures of computer networks'). The ontological approach is based on the use of abstraction operations PART_OF, IS_A, KIND -OF. They are semantic operations, which can be considered as a result of the abstraction creation method. Through the relation PART_OF it is possible to describe a set of features sufficient to pass an abstract object to the considered class. IS_A is a statement that a specific object that has certain values has been included in the same class. KIND-OF is a statement that specific objects enumerated by names were included in the considered class.

Using the above method of describing the concept as a unit of domain knowledge, it is important to distinguish the name of the concept, that is, the appropriate term, its definition and set of objects/phenomena that are subject to this concept within the limits of the chosen field. For example, thea term 'body' in the field of microbiology will be subject to a set of objects of a completely different nature than in the field of zoology or paleontology.

The etymological analysis of the term allows one to determine the ratio of the concept under study to other concepts or/and to establish a historical outline of its appearance. When developing curricula in a specific field, when a thesaurus is established, it is enough to use the names of concepts that belong to it. In the case of new fields, a lot of attention should be devoted to analysing the names of terms, which consists of several words.

For example, in the *e-Quality project*, the term 'education process' was examined in this respect. The use of this concept in a new, in contrast to the traditional, context related to the networked learning environment and the Bologna Process requires a proper definition of the concept.

Figure 4.9 lists three levels of aggregate features that describe the meaning of the teaching process at each level, and what's more, the hierarchy of levels allows you to follow the historical outline of the change in the volume of knowledge contained in the concept.

As an example of the development of the matrix concept structure in Table 4.2, the volume and depth of the *queuing system* concept in the field of mass service systems is presented, based on

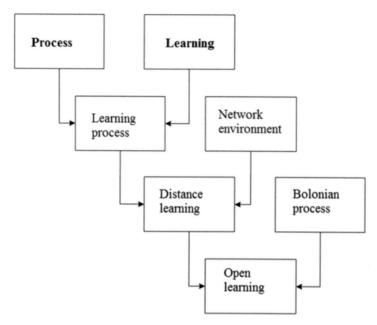

Figure 4.9 Historical outline of the concept: open learning.

the *Kendall notation* prepared for queuing systems. The presented matrix describes the concept of queuing systems, which are a kind of system of organising stochastic processes aimed at servicing a stream of repetitive orders/works.

The basic parameters of this type of systems are:

- Types of input streams (arrival pattern)
- Distribution of the service process (kind of servicing)
- Number of servers/jobs
- System capacity
- The potential number of system users (population)
- Discipline of servicing

A system of different combinations and element values of elements (servers, buffers, users, etc.) can be included in a given class of queuing systems:

Table 4.2 Matrix structure of concept 'queuing systems'

Queuing Systems	M/M/1	M/M/m	M/M/m/m	M/M/1/S	M/D/1/C	G/G/∞/prt	...
Arrival pattern	Markovian (Poisson) arriving process = $M(P)$	$M(P)$	$M(P)$	$M(P)$	$M(P)$	General	...
Kind of Servicing	Markovian (Exponential) servicing process = $M(E)$	$M(E)$	$M(E)$	$M(E)$	Deterministic	General	...
Number of servers	1	m	m	1	1	∞	...
System capacity	∞	∞	m	Limited by S	∞	∞	...
Population	∞	∞	∞	∞	Limited by C	∞	...
Discipline of servicing	First come first serve (FCFS)	FCFS	FCFS	FCFS	FCFS	Priority	...

- *M/M/*1: Markovian single-server system with infinite capacity, infinite population and service discipline – first arrived, first served (FCFS)
- *M/M/m*: Markovian multi-server system with infinite capacity, infinite population and FCFS support discipline
- Others according to Kendall's notation

Each column of the matrix contains a specification of the type of stochastic system under examination, which can be interpreted as a queuing system of a specific structure and organisation. The specification provided is sufficient to identify the appropriate mathematical model.

4.3 Ontological Scheme Formation for Knowledge Domain in ODL

This chapter describes a knowledge model oriented on the asynchronous mode of ODL. The formalisation of the knowledge model for a given domain, the operations on the knowledge and the algorithm of the knowledge model creation are submitted. All received decisions can be realised in a programme's environment compatible with the SCORM standard. The described methodology, based on a generalised knowledge model, enables developing an ODL learning course mainly for the fundamental knowledge. We describe the methodology and illustrate its use through a project to develop an ODL course for a queuing system. Moreover, a practical application is proposed based on the eQuality project (Zaikin *et al.*, 2004).

4.3.1 Concepts Network Creation Algorithm

The goal of the *concept network creation algorithm* (CNCA) (Fig. 4.10) is to identify the knowledge in a specific subject's domain and convert the knowledge to a form of a concept network (CN). The subject's domain's analysis is conducted on the first step of the expert's work. The subject matter expert analyses the domain, and the concepts are selected according to the rules of the concepts'

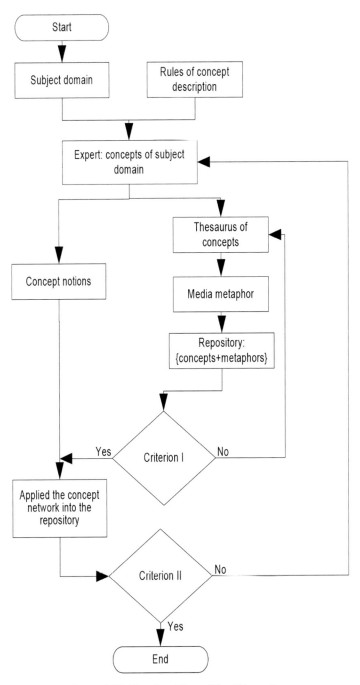

Figure 4.10 The algorithm of the CN creation.

descriptions. The identified set of concepts is enhanced by adding a semantic relation. Simultaneously, the creation of the concepts multimedia representation is being performed. The main function of each metaphor is to widely describe the concept or the idea which is covered by the concept. The abstract form of the concept implies its digital representation. The complete processing information, including concepts, meta-information (each notion is characterised by the author, data, etc.) media files, is stored in the repository. *Criterion I* of the CN's algorithm decides whether the choice of the multimedia metaphor is correct. The expert, who is carefully considering the judgement process, can enlarge the depth of the concept's metaphor based on his/her knowledge of the subject. After the positive semantic verification the CN is applied to the repository's structure. The whole system is evaluated again according to *criterion II*. The network of concepts should cover every bit of knowledge in the specified domain. The results of the algorithm are:

- CN,
- Repository,
- Hypermedia network of concepts

The last maps the concepts onto their multimedia representations that are saved in the repository. The algorithm's outcomes are fully compatible with the SCORM standard.

4.3.2 Didactic Materials Compilation Algorithm

The didactic materials' compilation algorithm (DMCA) (Fig. 4.11) adapts didactic materials for the student by taking under consideration the general educational standards and pedagogical conditions of the learning process. Generally speaking, the CN is adapted to a specific student by the content's personalisation. The goal of the algorithm is to develop a sequence of educational elements (in the sense of the SCORM 2004 standard), which is delivered to the student through the telecommunication network.

The CNCA creates a CN about the subject area. In the first step of the DMCA, the relationships in CN are ordered based on the student's

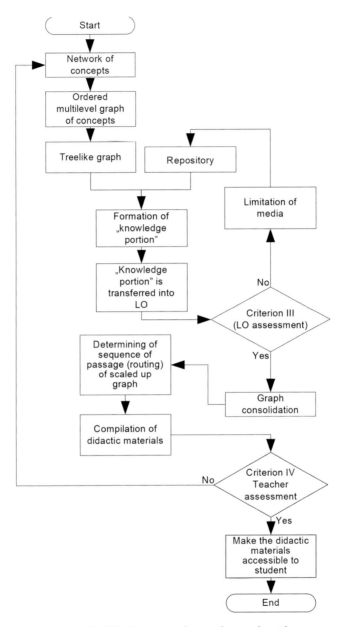

Figure 4.11 Didactic material compilation algorithm.

profile. In the next step, the ordered CN is transformed into a multi-level hierarchical graph with the basic concepts at the top level. Concepts at the low level of the hierarchical graph are interpreted as the objectives of learning. Such graphic representation of knowledge determines uniqueness of each concept and the number of its semantic links. The links define the semantic richness of each concept and its importance for other concepts.

At the next stage of the algorithm, the multi-level hierarchical network is decomposed into its own set of tree-like sub-graphs which are based on the methods described in this chapter. The trees are covered by a 'knowledge portion' mechanism, which is constructed based on cognitive regulations. The selected 'knowledge portion' is a basic element of the learning object. Transformation of the 'knowledge portion' into a learning object requires a representation of the concept in the repository content.

The data and its metadata are analysed based on *criterion III*. Criterion III estimates the 'size' of the data according to the telecommunication network's bandwidth limitation, the copyright issue and the resource accessibility. The clustering process joins the concepts of every 'knowledge portion' into one node. The knowledge incorporated into the 'knowledge portion' together with the essential metadata description creates the learning object. The generated learning objects are impinged on the CN, which is after that converted to a learning object's network. Afterwards the sequence of nodes (learning objects) is set. The established learning object's chain determines the way of learning for each student. Before the learning objects' sequence will be placed in the repository, all elements are converted to a SCORM-compatible form. The course instructor/teacher has to take the final decision about the didactic materials' acceptance (*criterion IV*). Finally, the didactic materials are accessible to every registered student.

4.3.3 Concepts Network's Relations

Similar to a semantic network, a CN does not attempt to represent the real mind's structure. It is no more than an abstract representation, which is an instrumental model adapted to the domain's knowledge manipulation in the form of ontology engineering. Researchers believe that knowledge in the mind is organised according to a cognitive economy: concepts with a higher number of semantic relationships are situated on the upper level of the knowledge representation's structure.

Generally speaking, a CN always covers knowledge from one specific domain and for that reason can be considered as an ontology. The statements confirming such a declaration come from previous assumptions.

The conceptual framework for a CN is based on the ontology characteristic. The concept ($\Phi = \langle X, T \rangle$) has been defined in the previous section. A *semantic relation* is defined as a relation between two concepts. Out of many other existing ones, the set of semantic relations $R = \{IS\_A, PART\_OF, \}$ is incorporated into the approach. According to (Goldstein and Storey, 1999) and (Storey, 1993) the inclusion, aggregation and association relations' classes are the most important ones. The last class notices some information about the association between concepts. In a CN this type of information is carried on by the inclusion and aggregation classes. The remaining relations' set R is, on the one hand, relatively small, but, on the other hand, the cohesion with the abstraction mechanism is preserved.

The *IS_A* relation, according to (Eynde and Gibbon, 2000), covers wide range of relations like: inheritance, implication and inclusion. The most popular *IS_A* application is taxonomy. If any elements of $T_1 = [\hat{t}_{ij}]$ (for the concept's name $N(\Phi_1)$) can be described as a matrix $T_2 = [\hat{t}_{ij}]$ (concept's name $N(\Phi_2)$) then the concept Φ_2 is connected with the Φ_1 by the *IS_A* relation.

The *PART_OF* relation is a base for several types of hierarchy (e.g. whole-part) (Sowa and Zachman, 1992). If $N(\Phi_1) \subset N(\Phi_3)$ and $T_1 = [\hat{t}_{ij}] \subset T_3 = [\hat{t}_{ij}]$, then the concept Φ_3 is connected to the Φ_1 by the *PART_OF* relation.

The ontology can be defined as a tuple:

$$\Omega = \langle S, \Pi \rangle, \tag{4.3}$$

where

$S = \{\Phi_1\}, i = 1,..., n$ is a set of concepts from a specific domain
$\Pi : S \times S \to R$ is a mapping from an ordered pair of concepts to the set of connections R (similar to (Câmara *et al.*, 2002))

In order to improve computer processing, the Ω form is transformed into a non-oriented graph G_Ω. The ontology in the form of the non-oriented graph G_Ω has the following definition:

$$G_\Omega = (V, E), \tag{4.4}$$

where

$V = \{v_i\}$ is a set of graph nodes ($i = 1, 2,..., n$), and every node matches one concept from the ontology.

$E = \{e_j\}$ is a set of graph edges ($j = 1, 2,..., m$). The edge between nodes v and w is defined as a symmetric relation $\{v, w\}$. The incidental matrix $A = \|a_{ij}\|$ corresponded to the graph G_Ω has the dimension $n \times n$ (where n is the total number of ontology concepts). Each element of the A matrix can be computed from:

$$a_{ij} = \begin{cases} 1, & \text{for } \{i, j\} \in E, \\ 0, & \text{if } \{ij\} \notin E. \end{cases} \tag{4.5}$$

The transformation ($\Omega \Rightarrow G_\Omega$) from Ω form to G_Ω form is defined in the following way:

$$PR_i = \begin{cases} 1, & \text{if } \Pi(s_i, s_j), \\ 0, & \text{otherwise.} \end{cases} \tag{4.6}$$

The reverse transformation ($G_\Omega \Rightarrow \Omega$) from G_Ω form to Ω form is defined in the following way:

$$\Pi(s_i, s_j) = \begin{cases} \aleph, & \text{if } a_{ij} = 1, \\ \emptyset, & \text{otherwise,} \end{cases} \tag{4.7}$$

where $\aleph$ is any element from the set: $\{IS\_A, PART\_OF, \emptyset\}$.

4.3.4 Scheme of the Concepts Network Creation Algorithm

This section presents some propositions for the CNCA implementation. The CNCA , briefly described in Section 4.3.1, can be put into service as a procedure shown in Fig. 4.12.

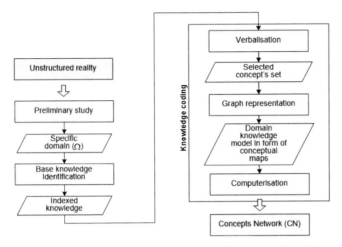

Figure 4.12 Scheme of the concept network creation algorithm.

In order to make the discussion of presented methods stricter, the queuing systems theory example is examined based on (Kleinrock, 1975; Ng, 1996). At the preliminary study step of the procedure unstructured reality is limited to the field of the discussed domain. The domain's boundaries have to be founded and specified. It will make creating the ontology possible. In case of the technical and mathematical sciences the existing taxonomies (e.g. the mathematics subject classification) provide some additional help, e.g. for the queuing system theory, the following association can be used:

> 60–xx Probability theory and stochastic processes
>> 60Kxx Special processes
>>> 60K25 Queueing theory

According to (Radosinski, 2001) on the basis of the knowledge identification step an expert decides what knowledge and information are useful in the ontology creation. The potential knowledge sources can be a related expert, corresponding ontologies and digital sources. After that, the knowledge is already indexed.

At the verbalisation step, the expert's knowledge is recorded. The verbalisation begins the knowledge coding process. The knowledge-coding process adapts knowledge to the requirements of the environment (e.g. LMSs/LCMSs). The knowledge in the DL environment is transformed into a collection of concepts Φ with their matrixes $T = [\hat{t}_{ij}]$ (see Table 4.2).

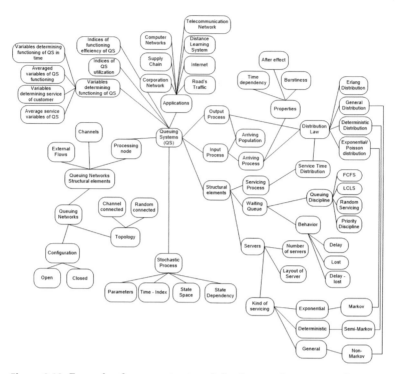

Figure 4.13 Example of a concept network for the queuing systems domain.

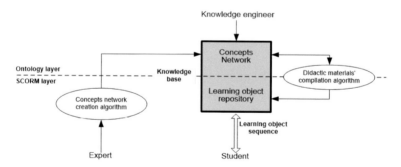

Figure 4.14 Scheme of knowledge base formation.

The knowledge is structured during the graph representation phase. This phase creates a network (e.g. conceptual maps), which plays the role of the formal knowledge representation model. During this phase the relations are set. An example of a CN for the queuing systems domain is shown in Fig. 4.13.

At the last step of the procedure, the CN is changed into a computer-ready form. The computerisation phase adjusts the structures of the ODL model to the digital form. The knowledge manipulation language defines the model's inputs and outputs. The final ontology can be expressed in one of the well-accepted onto-languages (e.g. FLogic (Benjamins *et al.*, 1999)).

4.3.5 Proposition for the Didactic Materials' Compilation Algorithm

The repository with the finished learning objects will be relatively large. A learning object is a strongly personalised piece of educational content, and because of this, each student should possess his/her own set of learning objects. On the basis of this assumption, the repository has to store an enormous number of learning objects in order to provide common access for many students, which is not feasible.

As (Ausubel, 1968; Hestenes, 1995) showed, the conceptual learning process is based on the concept manipulation process. The proposed idea is to store knowledge about a specific domain in an ontological, generic form. By taking advantage of the CN structure, knowledge can be stored in the repository on two layers, ontology and SCORM ones (Fig. 4.14). The ontology layer accumulates the knowledge in the CN form. On the basis of the CNCA, the expert creates the knowledge's domain model, where the CN's concept is specified and the depth matrix is established. On the second layer the knowledge is compiled into a SCORM-like structure. Nowadays almost any LMS/LCMS can import and operate with SCORM-like structures.

Every student has his/her own learning capabilities. Hence, the system should adapt the learning process to the individual student profile, which consists of information about the student's learning style (Liu and Ginther, 1999), previous knowledge and experience in the specific domain. Furthermore, the didactic content is affected by the learning objectives. On the basis of this information, the LMS/LCMS should adapt the CN to the learning object. The authors propose the DMCA to perform this task. The DMCA consists of several steps, which will be discussed next (more details about the algorithm can be found in (Różewski, 2004).

4.3.6 Concepts Network Dimension Reduction

The basic generic ontology (expert's ontology Ω_E) generated by the expert in many cases is beyond the learning objectives. The teacher, who carries in the mind the course intention, the student's profile and his/her characteristics, is obligated to set the course's aim and objectives. Therefore, the ontology has to be limited in order to achieve the teacher's expectations. In other words, the teacher's ontology Ω_T enters the expert's ontology Ω_E. The result's ontology $\Omega_{T(E)}$ (usually $\Omega_{T(E)} < \Omega_E$) covers the required student's course knowledge and is called a reduced expert's ontology $\Omega_{T(E)}$.

4.3.7 Basic Concepts Selection Using the Student's Profile

A set of concepts S_S represents the knowledge which has been learned and understood by the student in a specific domain, and it is defined in the student's profile. The learning process will be more effective if the relation between the acquired knowledge and the new one is emphasised. The common knowledge space allows the student to do knowledge assimilation. Therefore, such shared concepts can play the role of starting points for the new learning process.

The idea is to integrate the S_S set with the reduced expert's ontology $\Omega_{T(E)}$. For this, the basic concept's set should be obtained and the ontology $\Omega_{T(E)}$ has to be updated. If we move the discussion from the conceptual level to the mathematical one, the non-oriented ontological graph $G_\Omega = (V, E)$ should be converted into an oriented graph $\vec{G}_\Omega$, where the orientation is made on the basis of the S_S set. The supported transformation $\Omega \Rightarrow G_\Omega$ (from the abstract ontological form into the mathematical one) is discussed in Section 4.3.3.

In the queuing system example, the basic set of concepts S_S has the following form: $S_{S1} =$ {*Erlang Distribution, General Distribution, Deterministic Distribution, Exponential/Poisson Distribution, First Come First Serve (FCFS), Last Come Last Serve (LCLS), Random servicing, Priority Discipline*}.

The oriented graph $\vec{G}_\Omega$ provides additional information about the learning process. The student (teacher) can deduct the direction of learning by simply following the oriented relations. In Fig. 4.15, an

Figure 4.15 Queuing system concept network converted to an oriented graph using the S_{S1} set.

example of the oriented graph corresponding to Fig. 4.13 and S_{S1} set is represented.

The student begins with the basic concepts, which are included in the S_{S1} set, and then explores every concept according to the oriented relationship. The learning process finishes after the student has visited every concept (node of graph). The oriented relationship is made on the basis of the most likely direction the real learning process would occur.

The result of the discussed step is the oriented graph:

$$\vec{G}_{\Omega} = (V, \vec{E}),$$
(4.8)

where

$V = \{v_i\}$ is a set of nodes of the graph ($i = 1, 2,..., n$). Every node matches one concept from the ontology $\Omega_{T(E)}$.

$\vec{E} = \{e_j\}$ is a set of edges of the graph ($j = 1, 2, ..., m'$). The edge between nodes v and w is defined as a relation (v, w).

The incidental matrix $vecA = \|a_{ij}\|$ represents the oriented graph $\vec{G}_{\Omega}$. The matrix' elements can be computed from:

$$a_{ij} = \begin{cases} 1, & \text{for } \{i, j\} \in \vec{E}, \\ 0, & \text{for } \{ij\} \notin \vec{E}. \end{cases}$$
(4.9)

4.3.8 Hierarchically Ordered Concepts Network

The next phase of the DMCA transfers the oriented graph $\vec{G}_{\Omega}$ to the hierarchical ordered CN $\overline{\overline{G}}_{\Omega}$. Each concept of the hierarchically ordered network is characterised by a corresponding level. The network's nodes cannot be connected within the same level.

On the first level (i.e. level 1 in Fig. 4.16) the concepts which belong to the basic knowledge set S_S are located. These concepts are called basic ones. The other levels consist of the concepts included in the course's objective, and the student should memorise them (levels 2–10 in Fig. 4.16). The hierarchical relation gives information about relations between concepts.

The result of this phase is the hierarchically ordered CN which can be represented as a 3-tuple:

$$\overline{\overline{G}} = (L, \overline{\overline{V}}, \overline{\overline{E}}),$$
(4.10)

where

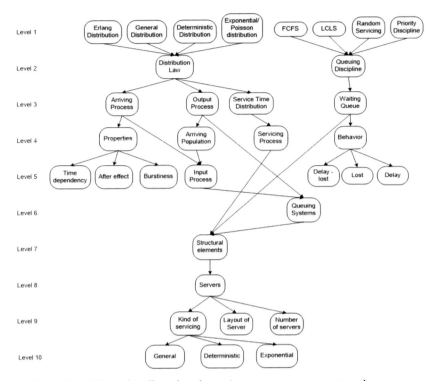

Figure 4.16 Hierarchically ordered queuing system concept network.

$L = \{l\}, l = 1, ...l^*$ is a set of levels

$\overline{\overline{V}} = \{v_{lk}\}$ is a set of network nodes, where l is the level's index and k is the current node's number on the level l

$\overline{\overline{E}} = \{e_i\}$ is a set of the graph's edges $(j = 1, 2,..., m', \overline{\overline{E}} = \vec{E})$

The incidental matrix $\overline{\overline{A}} = \|a_{ij}\|$ reflects the hierarchical network $(\vec{A} = \overline{\overline{A}})$.

4.3.9 Transformation of the Hierarchically Ordered Concepts Network

The main aim of the DMCA is the learning objects' sequence creation. In the DL environment rooted in the SCORM standard (SCORM 1.2 or better), the content's sequence is determined, based on the dedicated SCORM structure called activity tree (the SCORM standard is discussed in more details in Section 4.3.12. The next step of the DMCA converts the hierarchical network $\overline{\overline{G}}_\Omega$ into a forest $\widehat{\widehat{G}}_\Omega$. The forest consists of a set of trees $\widehat{G}_\Omega$. Every tree can be considered as a SCORM's activity tree. Finally, the set of activity trees makes up a course for LMS/LCMSs.

The number of trees in the forest depends on the learning method selected in the initial phase. For the inductive learning style the number of trees in the forest is equal to the number of leaves in the hierarchical network. On the contrary, in the deductive learning style the number of trees in the forest is identical to the number of nodes on the first level of the hierarchical network.

The result of the transformation of the hierarchically ordered CN into the forest $\widehat{\widehat{G}}_\Omega$ is the set of trees $\widehat{G}_\Omega$. Every tree has the following description:

$$\widehat{G}_\Omega = (L, \widehat{V}, \widehat{E}).\tag{4.11}$$

where
   $L = \{l\}, l = 1, ..., l^*$ is the level's set
   $\widehat{V} = \{v_{lk}\}$ is a set of graph nodes, where l is the level's index and k is the current node's number on the l level
   $\widehat{E} = \{e_j\}, j = 1, 2, ..., m$ is a set of graph edges

During the transformation $\overline{\overline{G}}_\Omega \rightarrow \widehat{\widehat{G}}_\Omega$ a duplication problem occurs. Moving from the hierarchically ordered CN into the set of trees causes some concept duplication. The operation of duplication indicates that some nodes of the hierarchically ordered CN will be doubled in order to get the tree-like structure. The duplication operation guarantees that every concept will be present in the comprehensive context, which ensures a high level of accuracy.

4.3.10 Creation of Overlapping Portions

One learning object m which includes just one concept is not a feasible ratio for an efficient work. It is easy to imagine a tree with numerous nodes, which are difficult to maintain. Moreover, from the organisation's point of view such a structure will also cause administration and security difficulties. The authors came up with a new idea. In the presented approach, learning objects accumulate similar concepts into one object. In the next step, the tree $\widehat{G}_\Omega$ will be covered with the portions, which will be further converted into a learning object. A portion is a sub-graph of the initial tree graph $\widehat{G}_\Omega$. All portions of the discussed tree are included in the portion's set $g_p = \{\widehat{G}_p^i\}$, where $i = 1,...,p^*$.

The portion's $\widehat{G}_p$ definition is as follows:

$$\widehat{G}_p = (L_p, \widehat{V}_p, \widehat{E}_p), \tag{4.12}$$

where

$L_p = \{l\}$ is a level's set for the portion $\widehat{G}_p$, $L_p \subset L$
$\widehat{V}_p = \{v_i^p\}$ is a node's set for the portion $\widehat{G}_p$, $\widehat{V}_p \subset \widehat{V}$
$\widehat{E}_p = \{e_i^p\}$ is an edge's set for the portion $\widehat{G}_p$

A portion's creation rules are as follows:

(1) Every portion $\widehat{G}_p$ is a sub-graph of $\widehat{G}_\Omega$, $\widehat{A}_p \subset \widehat{A}$.
(2) The sum of the sub-graphs $\widehat{G}_p$ gives the $\widehat{G}_\Omega$ graph, in which the portions unrelated to $\widehat{G}_\Omega$ cannot be allowed.
(3) Every portion $\widehat{G}_p$ has a limited number of nodes.
(4) Every portion has to contain at least one node common with another portion, i.e. $\widehat{G}_p^i \cap \widehat{G}_p^j \neq \emptyset$, where $\widehat{G}_p^i, \widehat{G}_p^j \in g_p, i \neq j$.

As we already mentioned before, learning objects accumulate similar concepts into one object. The question is, how many concepts should be incorporated into one learning object? The answer comes from cognitive science. Research on the mind provides a useful memory model, which consists of the following memory types: sensor memory, short-term (ST) memory and long-term memory. The information from the environment is converted into an electrical signal by the sensor memory. After that, on the basis of the knowledge obtained from the unlimited long-term memory, the information is analysed in ST memory. Finally, some part of information is encoded in ST memory in the form of a cognitive structure and then sent to the long-term memory as knowledge.

The most important type for our study is ST memory, due to its limited capacity. The idea is to relate the 'size' of the learning object to the ST memory capacity. The ST memory capacity is determined by the Miller Magic number (7 ± 2) (Miller, 1956), which was updated

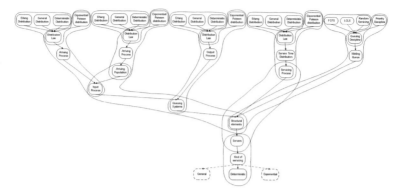

Figure 4.17 Tree with portions ($P_{max} = 3$).

by Cowan (about 4) (Cowan, 2001). The newest research (Gobet and Clarkson, 2004) brings up a new estimation (about 3). Another study (Visser *et al.*, 1997; Vogel and Machizawa, 2004) is focused on the empirical experiments and localised ST memory as a tiny spot in the posterior parietal cortex.

An additional significant indicator is the unique nature of the ST memory capacity. The limitation is counted by chunks, not bites. A chunk (Gobet *et al.*, 2001; Gobet and Simon, 2000; Simon, 1974) is a set of concepts which are strongly related to each other and, at the same time, weakly related with other concepts. In the case of a concept's definition from Section 4.2.2, a concept is consider as a chunk. Therefore, the number of concepts in $\widehat{G}_p$ should not overcome P_{max}, where P_{max} is the ST memory capacity set by a psychological survey. In Fig. 4.17 is shown an example of a tree, which was obtained from the hierarchical ordered queuing system CN (Fig. 4.16).

As a result, the discussed step generated a set g_p, which is a subset of $\widehat{G}_p$ for every tree in the forest $\widehat{G}_\Omega$.

4.3.11 Graph's Clustering

The last phase of the DMCA is the graph's clustering. In this step any portions of any tree $\widehat{G}_\Omega$ from the forest $\widehat{G}_\Omega$ are integrated into the form of a consolidated node W_S. The consolidated node has a name, which comes for example, the largest integrated concept. The clustered graph S has the following description:

$$S = (N_S, W_S, E_S), \qquad (4.13)$$

where

$N = \{n\}$ is a set of levels for the clustered graph S

$W_S = \{w_i\}$ is a set of nodes for the clustered graph S, $i = 1,..., card(g_p)$

$E_S = \{e_i\}$ is a set of edges for the clustered graph S

The clustered graph S can be converted into the SCORM activity tree structure. The sequence of learning objects in the SCORM nomenclature is called an activity tree. The clustered graph obtained from the tree (Fig. 4.17) is demonstrated in Fig. 4.18. The clustered graph is converted to a SCORM activity tree structure (shown in Fig 4.19). On the basis of the activity tree, the LMSs/LCMSs generate a learning object sequence. In general, LMSs/LCMSs use a pre-order strategy. In Fig. 4.19 the sequence is as follows: A, B, BA, BAA, BB, BBA, BC, BCA, BCAA, BCB, BCBA, BCC and BCCA. Moreover, the SCORM standard can use a script language in order to prevent repetition (e.g. distributed law in Fig. 4.15). The built-in script language gives advanced sequence control options. For instance, the following rule *IF node 4. state=completed THEN go .previous* can be applied.

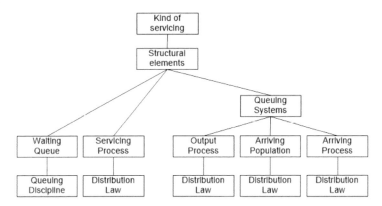

Figure 4.18 Clustered graph.

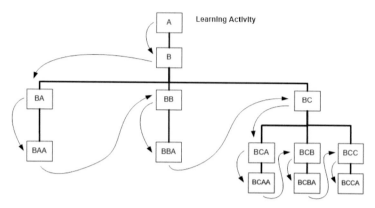

Figure 4.19 Activity tree based on the clustered graph from Fig. 4.18.

4.3.12 SCORM Framework

The SCORM layer model consists of four conceptual layers (Fig. 4.20). On the bottom is the data layer (1). The data layer includes all data files, which are used to the concept's metaphor creation. Furthermore, the data layer contains all metadata files used at higher layers of the SCORM model (i.e. as a part of the IEEE LOM declaration for the Sharable Content Object). The second layer (2) covers the structures intended to data integration. There are a number of SCORM integrated structures: Sharable Content Object, Asset, Content Aggregation. Each of them has its own, special role. The Content Aggregation structure is the most important one due to integration features. The Sharable Content Object and Asset can be integrated into a learning object-like structure. The third layer (3) is a sequence layer. On this level the transformation Content Aggregation

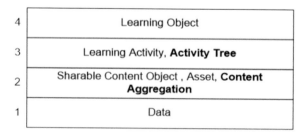

Figure 4.20 SCORM layer model.

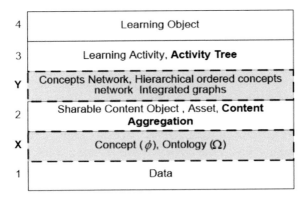

Figure 4.21 Enhanced SCORM layer model.

→ Activity Tree occurs. Simultaneously, the transformation Asset, Sharable Content Object → Learning Activity takes place and the learning object's content is made. On the highest layer (4) the mechanisms of LMS/LCMS create the learning object based on the activity tree analysis.

The goal of the proposed algorithms is to enhance the SCORM standard by adding two additional layers (X, Y). The CNCA is implemented between layers 1 and 2 (layer X in Fig. 4.20). The CNCA layer works with the ontology and concept structures. Concepts can be seen as Sharable Content Objects. The matrix of the concept's depth expressed in XML is adapted to the SCORM standard. Moreover, the ontology can be considered a generic form of the SCORM Content Aggregation.

Another proposed algorithm, i.e. the DMCA, enters the second additional layer. This layer is replaced between layers 2 and 3 . The main task of layer Y is to adapt the CN to the requirements of SCORM sequencing theory. Based on the proposed algorithm, the network ontology structure is transformed to the forest. Each of the forest's trees has the nature of an activity tree, and the tree's nodes are the learning activities (Fig. 4.21).

4.3.13 Application

The quality issue is a difficult problem, especially in human-centred technology (Sharples *et al.*, 2002). Around the world, quality research projects are running in order to find a methodology or procedure that allows achieving a high level of quality. One of the biggest issues is the quality in ODL (McGorry, 2003). Therefore, a deep investigation is required. The authors have done some research in the framework of the European eQuality project ("e-Quality", 2004; Zaikin *et al.*, 2004).

The eQuality project objectives are to produce a common reference methodological framework for quality in ODL in Europe, taking into account cultural differences, experiences, needs and best practices, joined with the ongoing normalisation work. The project incorporates several European universities: European Universitary Pole of Montpellier and Languedoc – Roussillon (France), University Montpellier 2 (France), Open University of Catalonia (Spain), University of Tampere (Finland), Technical University of Szczecin (Poland), University of Applied Sciences Valais (Switzerland) and Lausanne University (Switzerland).

The algorithms (CNCA, DNCA) presented in this chapter are being successfully applied in a training session organised to provide feedback on the eQuality project's methodology and tools. The training session is set in the Simulation Web Portal (SWP) environment (Fig. 4.22). The SWP consist of four main modules/sections: simulation principles, online tests, online simulation and personal files management (Kushtina and Różewski, 2004). The first section is a simple textbook on computer simulations. Its aim is to introduce

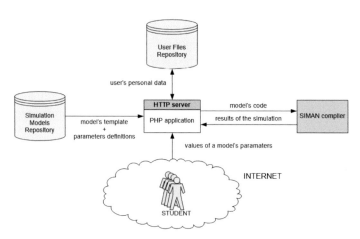

Figure 4.22 Architecture of the online simulator.

students to the computer simulation principles. The second section is a set of exercises where students can verify their knowledge. In order to solve some of the exercises, students are asked to build a simple simulation model, carry out a simulation experiment and give its results. These tasks can be done in the next section 'online simulation'. The online simulator is based on the SIMAN language compiler from Rockwell's ARENA simulation package. This system is highly specialised in serving educational purposes and was designed to demonstrate the abilities and advantages of computer simulation. In the last section, 'personal files management', a registered user can save and open simulation results and parameters of all models developed in the 'online simulation' section (Chapter 8).

4.3.14 Summary

1. The proposed approach provides a high quality of interaction between all participants of the distance learning process, namely:
   - More precisely shares functions of the expert of the subject domain and the teacher which prepares methodical materials for distance learning
   - Allows to carry out an exchange between the student and the teacher at the level of knowledge
2. The model of knowledge representation and transfer can serve as a platform for cooperation between the teacher and the expert. The new teacher's role in distance learning provokes a more creative approach to the analysis of the conceptual scheme of the subject domain and the choice of a teaching method, depending on the objectives of learning and the initial knowledge of students.
3. The developed algorithms allow uniting the knowledge of the subject's domain and the methods of teaching, thus providing high-quality methodical materials. Apart from that, algorithms can also be used as a methodical toolkit for the teacher, working in both traditional and distance learning systems.

4.4 Case Study 1. Ontological Model of Relational Database of Intangible Production

The Concept of ontological model of database

The basic objectives of an ontological *relational data model* (RDM) are to present a formalised interface between specialists in a specific domain and specialists in database design. An RDM is an exhaustive data model needed for implementation of all functions of the information management system. The ontological model of the database consists of two parts: information objects (IOs) and structural relations between the IOs.

Information objects

An IO reflects a real object, process, event or phenomenon. Qualitative and numerical characteristics of IOs change as a result of real processes carried out in the production process (PP) of the enterprise. A description of IOs in the form of a set of attributes enters the RDM. An *attribute* is a logical, non-divided data element corresponding to a certain property of the object. An attribute is characterised by an *attribute schema,* , which includes:

- Name, type and class of the attribute value
- Restrictions on possible attribute values (format, unit of measurement, accuracy)
- Special properties (attribute security key, control of attribute value trust)

Depending on the possible value, attributes can be divided into the following types: numeric, text, symbolic, logic and special purpose (e.g. temperature, date, coordinates). The *attribute template* specifies the format (size) and type of the value.

Here are the following attribute templates:

- *(9n)* is an integer value of *n* numbers
- *s 9 (n) .9 (m)* is the real value (*n, m*), *s* is thesign of the attribute,
- *X(p)* is the symbolic value, *p* is the number of symbols in the attribute value
- *A(p)* is the text value (Cyrillic or Latino)

Example:

- Date of birth *9 (6)* : xx xx xx (day, month, year),
- Air temperature *s. 9 (2) .9 (1)*
- Product code *x (16)*

Depending on the nature of the indicated property all attributes in the IO can be divided into qualities, that reflect the qualitative properties of the object, and bases,which reflect the numerical characteristics of the object.

- *Attributes - the qualities* used for the logical processing of information (e.g. sorting, grouping, searching, calling, etc.)
- *Attributes - bases* expressed as an integer or real properties of objects

Example: Specify the attribute roles in the document 'Material Loading Card in Stock' (Table 4.3).

   Attributes can be associative , forming a *complex information unit* (CIU). A CIU is exhaustive characteristics of the IO. A CIU can only consist of attributes or other CIUs.

Examples of CIUs

- *Techno-economic indicator* – consists of one attribute-base and several attribute-qualities, characterising the attribute-base.
- *The document* is called the material object that contains in a fixed form the information formed in a fixed order and which has a legal basis. The document is a complex CIU.

The graphical form of a CIU can be presented in the form of a hierarchical graph. The vertices of the graph are attributes and other CIUs. The edge of the graph is the hierarchical relation between the attributes (CIUs).

Document example: Material loading card on stock

This document corresponds to the graph (Fig 4.23)
In Fig. 4.23 the symbol $S1$ marks the document and symbols $C11, C12, C13$ and $C14$ mark four CIUs.

Table 4.3 Attributes of the document 'Material Loading Card in Stock'

Attribute name	Attribute type	Template	Property qualitative or numerical
Number of the card	Numerical	x(6)	Identification of the object (primary key)
Date of completion	Numerical	9(3)	Secondary key
Kind of operation	Coding	x(3)	Qualitative
Store recipient	Coding	x(3)	Qualitative
Sender's plant	Coding	x(3)	Qualitative
Material name	Text	A(30)	Qualitative
Material code	Coding	x(3)	Qualitative
Unit	Text	A(20)	Qualitative
Amount sent back	Numerical	9(6)	Numerical
Amount accepted	Numerical	9(6)	Numerical
Price of the material	Numerical	9(6), 9(2)	Numerical
Summa	Numerical	9(6)	Numerical

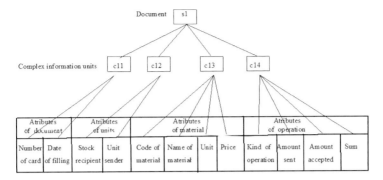

Figure 4.23 Graphical form of a complex information unit.

Analytical form of a CIU: $s(C11.(P1, P2), C12.(P3, P4), C13 \cdots)$ The dot is the sign of the hierarchical relationship. The comma is the character of the relative sequence within one hierarchical level.

A CIU is single if it is a unit value of a data and a massif if it contains multi-values of data.

Among the attribute-bases of a CIU, the most important is the key attribute (key) allowing in a unique (one-to-one) way to identify a copy of the CIU.

4.4.1 Requirements of the CIU Structure (3-Normal Form)

1. *Full functional dependence (FFD) of non-key attributes on the key attribute value*

   FFD: Let attributes A and B enter the IO. Then attribute B is in FFD on attribute A if at any time one value of attribute A corresponds to no more than one value of attribute B.

   Example: Material Code - Material Price

2. *Mutual non-dependency of key attributes (if the key consist of a few attributes)*

   If the key attribute of the IO is complex, e.g. two-component (A, B) then the non-key attribute C is in FFD on the key (A, B) if at any moment in time one attribute value (a, b) corresponds to no more one value of attribute c.

   Example: (*Material Code, Supplier Code*) - (Delivery Number).

The attribute 'code of material' is the primary key, and the attribute 'supplier code' is the secondary key. Both keys are independent attributes. There may be three or more key attributes in the IO.

Example: (*Material Code, Supplier Code, Delivery Number*) - Delivery Date, Delivery Number

3. *Mutual non-dependence of non-key attributes, which means no transitive relations between them*

Example: (*Material Code*) - (Material Price, Material Name, Number at the stock). All non-key elements are independent attributes.

Example of a complex information unit

Let the delivery be determined one-to-one by the pair (*Material Code, Supplier Code*).

The first requirement (FFD) needs to divide all attributes into three groups:

- Attributes that functionally depend on the primary key *Material code*: *Price of material, Name of material*
- Attributes that functionally depend on the secondary key *Supplier code*: *Supplier Name*
- Attributes that functionally depend on a pair of key attributes *Material code, Supplier code*: *Number of Delivery, Total Delivery, Delivery Date*

Therefore we received three IOs:

- *Material Material Code*: Material Price, Material Name
- *Supplier Supplier Code*: Supplier Name
- *Delivery Material Code, Supplier Code*: Delivery Number, Delivery Total, Delivery Date

4.4.2 Structural Relations of Information Objects

A relationship is a combination of structural relations and function links.

1) *Structural relationships between IOs*

A structural relation is defined between two IOs at the level of their items and is a binary relation. There are four types of structural relationships:

a) *Relation 'one-to-one* $(1 : 1)$

$$A - B$$

Two IOs A and B are in a one-to-one relationship, $1 : 1$ if every item A at every moment corresponds to one and only one item of object B and vice versa.

b) *Relation 'one-to-many* $(1 : M)$

$$A \longrightarrow B$$

Two IOs A and B are in a *one-to-many relationship* $1 : M$ if each item of object A at each time point corresponds to several items of object B and every item of object B at every time point corresponds to one and only one copy of object A.

c) *Relation 'many-to-one* $(M : 1)$

$$A \longleftarrow B$$

Two IOs A and B are in many-to-one relationship $M : 1$ if each item of object A at each time point corresponds to one and only one item of object B and every item of object B at every time point corresponds to several items of object A.

d) *Relation 'many-to-many* $(M : M)$

$$A \longleftrightarrow B$$

Two IOs A and B are in a many-to-many relationship $M : M$ if each item of object A at each time point corresponds to several items of object B and every item of object B at every time point corresponds to several items of object A.

Example: The RDM Material-Supplier-Delivery is represented in Fig. 4.24.

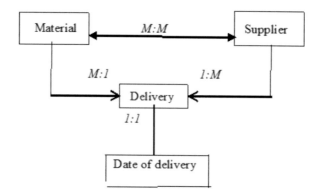

Figure 4.24 Relational data model 'material-supplier-delivery'.

Figure 4.25 Example of structural relations and functional links.

2) *Functional link between IOs*

   Functional links represent algorithms for processing and searching information. Functional links are an element of the information search algorithm, which shows the order in which the objects of the IOs are selected in the processing algorithm. Inputs of function links are source IOs. Outputs of function links are final IOs. There are two types of function links: *simple functional link*, which has only two IOs and a *complex functional link*, involving three or more IOs.

Example

Let's have the following query: After given a material code you have to find the suppliers of this material and specify all material deliveries. This query can be accomplished using two simple function links:

- Material (input object) / Delivery (output object) → M/D
- Delivery (input object) / Supplier (output object) → D/S

The functional link is a combination of two simple links in one algorithm that graphically is shown in Fig. 4.25. A relationship is a combination of structural relations and functional links.

4.4.3 Example of Relational Database Model Development

The selected company is a printing company dealing in intangible production (IP) (e-books, multimedia presentations, websites, etc.). The created information system (IS) is to help manage (including control and supervise) workflows, in addition to collecting information about orders and customers, manufacturing products and tools used in the enterprise to manufacture products. Typical offers of printing services:

- Design
- Digital processing, scanning
- Computer composition
- Image processing, printing
- Binding

Examples of products: leaflets, posters, calendars, notebooks, catalogues, books, letters, etc.

A number of IOs in a domain area is represented in Table 4.4.

The identification of types of relationships between IOs is represented in Table 4.5. A graphical representation of the matrix of relationships is represented in Fig 4.26.There four types of relationships between objects:

- $1:1$ – one-to-one relationship
- $1:M$ – one-to-many relationship
- $M:1$ – many-to-one relationship
- $M:M$ – many-to-many relationship

Modern database management systems perform only the first three types of relationships and do not implement the $M:M$ relationship.

Table 4.4 A set of information objects

	Name of the object	Semantics of an object	Marking of an object	Minimal composition of attributes
1	Parts	Goods Details Semi-products Materials	CZE (PRO)	*Part code* Part name Part Type Unit Code Price
2	Providers	Companies supplying materials and parts under cooperation	DOS	*Supplier Code* Supplier Name Address of Supplier
3	Consumers	Organisations and enterprises that receive ready productions	KON	*Consumer Code* Consumer name Address Consumer
4	Technology Group	Lots of machines of one type	GTE	*Group code* Group name Number of machines in
5	Professions and Competencies	Professions and Competencies needed in the production process	PKO	*Profession code* Name of profession Number of laborers. a profession
6	Structured Departments	departments sections sectors	DZI	*Department Code* Department Name Control Body
7	Personnel working	Staff working at the enterprise	PRA	*Employee ID* First name and last name Workplace
8	Machines	Machinery workshops Devices	MAS	*Inventory number* Time fund Expense
9	Delivery	The process of planned and actual delivery of materials and parts after cooperation	DWA	*Supplier Code* *Part (material) code* Delivery volume — planning: — real Delivery date — planning: — real
10	Order	The process of delivering the goods	ZAM	*Consumer Code* *Product code* The number of goods delivered Date of delivery
11	Specification of the product (parts).	Single-level composition The standard of using materials in part	SPT	*Product code* *Code of the entering part* Number of entries Usage standard
12	Technological route	Sequence of technological operations for the production of parts	MTE	*Part code* *Department code* *Technological group code* *The number of the operation* Profession code Production norms
13	Organisational structure	Direct composition and subordinate organisational units	STO	*Top division code* *Subordinate unit code* Number of the regulation
14	Measurement unit		JEM	*Unit code* Unit name

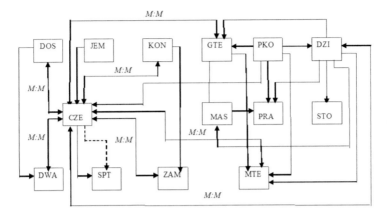

Figure 4.26 Graphical representation of the matrix of relationships.

For the implementation of a many-to-many relationship $M : M$, we need a *complex object-connection*. The minimal composition of attributes for the object-connection is the key of these objects, which is related by the relation *M: M + attributes*, values of which uniquely define the value of the component key. A complex object-connection can be a natural object, existing in the real world or artificial, created to realise a $M : M$ relation.

4.4.4 A Complex Information Object as a Connection between Simple Objects

Example 1
Let's consider a $M : M$ relationship between Consumer and Product:

This relationship can be realised by introducing a natural object Order: *Consumer, Product*, Atr1 = Order number, Atr 2 = Order Date, ...
This object consists of the key associated with the object and non-key attributes that uniquely depend on the pair of keys *Consumer, Product*.

Table 4.5 Matrix of relationships between information objects

	CZE	DOS	KON	GTE	PKO	DZI	PRA	MAS	DWA	ZAM	SPT	MTE	STO	JEM
CZE	—	$M:M$	$M:M$	$M:M$	$M:M$	$M:M$	—	—	$1:M$	$1:M$	$1:M$	$1:M$	—	$M:1$
DOS	$M:M$	—	—	—	—	—	—	—	$1:M$	—	—	—	—	—
KON	$M:M$	—	—	—	—	—	—	—	—	$1:M$	—	—	—	—
GTE	$M:M$	—	—	—	$M:M$	$M:M$	—	$1:M$	—	—	—	$1:M$	—	—
PKO	$M:M$	—	—	$M:M$	—	$M:M$	$1:M$	—	—	—	—	$1:M$	—	—
DZI	$M:M$	—	—	$M:M$	$M:M$	—	$1:M$	$1:M$	—	—	—	$1:M$	$1:M$	—
PRA	—	—	—	—	$M:1$	$M:1$	—	$M:1$	—	—	—	—	—	—
MAS	—	—	—	$M:1$	—	$M:1$	—	—	—	—	—	—	—	—
DWA	$M:1$	$M:1$	—	—	—	—	—	—	—	—	—	—	—	—
ZAM	$M:1$	—	$M:1$	—	—	—	—	—	—	—	—	—	—	—
SPT	$M:1$	—	—	—	—	—	—	—	—	—	—	—	—	—
MTE	$M:1$	—	—	$M:1$	$M:1$	$M:1$	—	—	—	—	—	—	—	—
STO	—	—	—	—	—	$M:1$	—	—	—	—	—	—	—	—
JEM	$1:M$	—	—	—	—	—	—	—	—	—	—	—	—	—

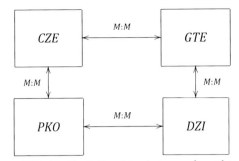

Figure 4.27 $M : M$ relationships between four objects.

Such attributes are, for example, Order Number and Order Date:

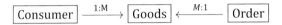

Example 2

Object - a relationship can transform $M : M$ relationships between three and more objects,

Let's look at all the $M : M$ relationships between: *the Part (CZE), Technology Groups (GTE), Professions and Competencies (PKO) and Divisions (DZI)* (Fig 4.27).

Four $M : M$ relationships can be eliminated by introducing a natural object.

Technology Route (MTE): (*Part Code*, *Division Code*, *Technological Group Code*, *Professional Code*, *Number of operations*, *Production norms*)

This connection-object consists of the keys of related objects and non-key attributes that uniquely depend on the four keys. These attributes are (*Operation Number, Production Norms*).

From the graphics it is apparent that all $M : M$ relationships are redundant and can be implemented using the MTE-Connection object. So we can eliminate all $M : M$ relationships between objects, listing them for relationships of type $1 : M$, as shown in Fig 4.28.

If there is no natural connection-object between objects $A\{code1,...\}$ and $B\{code2,...\}$, then they will introduce an artificial object $C : \{code1, code2\}$.

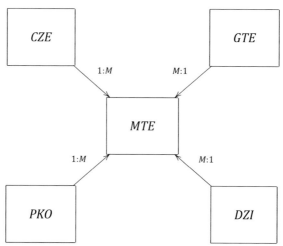

Figure 4.28 Transformation of $M : M$ relationships between four objects.

4.4.5 Elimination of Redundant $M : M$ Relations between Information Objects in the Matrix of Relationships

The analysis of the relationships matrix indicates that all relations of the type $M : M$ are redundant because they can be implemented by the relations of type $1 : M$ (Table 4.6).

A graphical representation of the matrix of the transformed matrix of relationships is represented in Fig. 4.29.

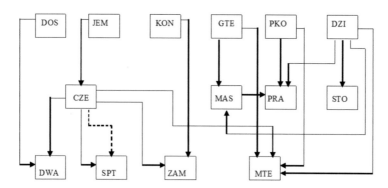

Figure 4.29 Graphical representation of the transformed matrix of relationships.

Table 4.6 Matrix of relationships between information objects without $M : M$ relations

	CZE	DOS	KON	GTE	PKO	DZI	PRA	MAS	DWA	ZAM	SPT	MTE	STO	JEM
CZE	—	X	X	X	X	X	—	—	1 : M	1 : M	1 : M	1 : M	—	M : 1
DOS	X	—	—	—	—	—	—	—	1 : M	—	—	—	—	—
KON	X	—	—	—	—	—	—	—	—	1 : M	—	—	—	—
GTE	X	—	—	—	X	X	—	1 : M	1 : M	—	—	1 : M	—	—
PKO	X	—	—	X	—	X	1 : M	—	—	—	—	1 : M	—	—
DZI	X	—	—	X	X	—	1 : M	1 : M	—	—	—	1 : M	1 : M	—
PRA	—	—	—	—	M : 1	M : 1	—	M : 1	—	—	—	—	—	—
MAS	—	—	—	M : 1	—	M : 1	M : 1	—	—	—	—	—	—	—
DWA	M : 1	M : 1	—	M : 1	—	—	—	—	—	—	—	—	—	—
ZAM	M : 1	—	M : 1	—	—	—	—	—	—	—	—	—	—	—
SPT	M : 1	—	—	—	—	—	—	—	—	—	—	—	—	—
MTE	M : 1	—	—	M : 1	M : 1	M : 1	—	—	—	—	—	—	—	—
STO	—	—	—	—	—	M : 1	—	—	—	—	—	—	—	—
JEM	1 : M	—	—	—	—	—	—	—	—	—	—	—	—	—

4.4.6 Canonical Form of the Relational Model

A relational model of a domain is presented in *canonical form* if four conditions are followed:

1) The IOs are structured correctly and presented in a 3D normal form.
   - The FFD of non-key attributes on the key attribute value
   - Mutual non-dependency of key attributes (if the key is complex)
   - Mutual non-dependence of non-key attributes
2) There are transformed $M : M$ relationships used in simple functional links.
3) IOs are arranged in hierarchical levels, i.e. single arrows $(1 : M)$ are oriented to the bottom.
4) A complex functional link is transformed to a set of simple functional links.

Example: The complex function link $(A, B/C)$ can be converted into two simple function links (A/C) and (C/B), as shown in Fig. 4.30.
   The placement of IOs information objects at hierarchical levels can be accomplished by processing adjacent matrices. The adjacent matrix is a square matrix in which the number of rows and columns is equal to the number of IOs.

$$A = \|a_{ij}\|, \quad i = 1, \dots n, \quad j = 1, \dots n, \quad a_{ij} = \{_0^1 \qquad (4.14)$$

$$(A, B/C) \Rightarrow (A/C) + (C/B)$$

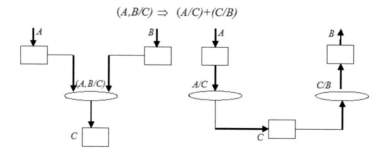

Figure 4.30 The transformation of complex function link $(A, B/C) \Rightarrow (A/C) + (C/B)$.

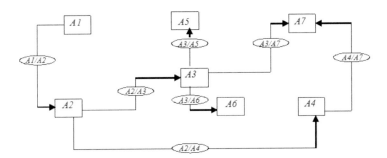

Figure 4.31 Graph of relationships and functional links.

The binary element of the matrix $a_{ij} = 1$, if the IOs o_i and o_j, which are appropriate in the $i - th$ line and in the $j - th$ row, are connected by the relation $1 : M$. Otherwise $a_{ij} = 0$.

Example
Let us consider the following graph of relationships and functional links (Fig. 4.31).
Task: Rearrange the objects of the graph of relationships, corresponding to matrix A_i at the hierarchy levels, so that the graph is tree-like (at the top of the tree are *rooted objects* and at the bottom are *leaves of the tree*).
We introduce the *adjacent matrix*:

$$P = \|p_{ij}\|,$$

where the number of rows and columns is equal to the number of IOs

$$i, j = A_1, A_2, ..., A_k.$$

Element of the matrix

$$p_{ij} = \begin{cases} 1, & if\_the\_object\ A_i\_in\_relation\_1 : M\_with\_the\_object\_A_j \\ 0, & \_otherwise \end{cases}$$

The objects A_i and A_j are connected by a functional link.
For a given example, the matrix P is as follows (Table 4.7).

Table 4.7 The adjacent matrix for the graph

	A_1	A_2	A_3	A_4	A_5	A_6	A_7
A_1		1					
A_2			1	1			
A_3					1	1	1
A_4							1
A_5							
A_6							
A_7							

4.4.7 Algorithm of Processing Adjacent Matrix $P = \|p_{ij}\|$

1. We sum the elements of the matrix in each column.

	A_1	A_2	A_3	A_4	A_5	A_6	A_7
Sum	0	1	1	1	1	1	2

2. Column analysis, searching for columns with value 0,
   $A_1 = 0$.

3. In matrix P we draw the column and corresponding row A_1, and again we sum the elements of the remaining columns.

	A_1	A_2	A_3	A_4	A_5	A_6	A_7
A_2	X		1	1			
A_3	X				1	1	1
A_4	X						1
A_5	X						
A_6	X						
A_7	X						
Sum	X	0	1	1	1	1	2

4. Column analysis, searching for columns with value 0,
   $A_2 = 0$

5. In matrix P we draw column A_2 and its corresponding row A_2 and again we sum the elements of the remaining columns.

6. Column analysis, searching columns with value 0,
   $A_3 = 0, A_4 = 0$.

	A_1	A_2	A_3	A_4	A_5	A_6	A_7
A_3	X	X			1	1	1
A_4	X	X					1
A_5	X	X					
A_6	X	X					
A_7	X	X					
Sum	X	X	0	0	1	1	2

7. In matrix P we draw columns A_3, A_4 and the corresponding lines A_3, A_4 and again we sum the elements of the remaining columns.

	A_1	A_2	A_3	A_4	A_5	A_6	A_7
A_5	X	X	X	X			
A_6	X	X	X	X			
A_7	X	X	X	X			
Sum	X	X	X	X	0	0	0

8. Column analysis, searching for columns with a value of 0,
   $A_5 = 0, A_6 = 0, A_7 = 0$
   We re-build the graph of relationships and functional links (Fig. 4.32).

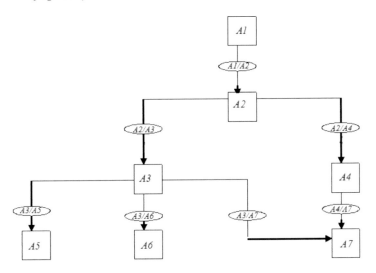

Figure 4.32 Re-build graph of relationships and functional links.

4.4.8 Example of the Project "Develop a Relational Database Model in the Given Subject Area"

Formulation of the project

I Tasks for a specialist in the domain area
- Select the subject area and give a description of the domain (e.g. production company).
- Specify the IOs of the domain.
- Identify sets of key and non-key attributes of each IO.
- Specify the characteristics of each object and the attribute of the object in tabular form.

II Tasks for system analyst
- Prove that the structure of each IO meets the requirements of the third normal form.
- Determine a matrix of relationships between objects and build a graph of relationships.
- To analyse and eliminate relationships of type M:M, replace them into relationships of type $1 : M$ by introducing new natural or artificial IOs.
- Identify key and non-key attributes of new objects.
- Provide a modified matrix and a graph of relationships.
- Describe the semantics and characteristics of relationships in tabular form.
- Develop a canonical structure of the relational model of the domain using the hierarchical graphing algorithm.

III Tasks for a specialist in software engineering
- Formulate and describe typical inquiries in the domain area and determine their relevant functional link.
- Formulate common queries in the data manipulation language (SQL type).

Table 4.8 A set of information objects

N	Name of object	Notation
1	Products (R)	PRD
2	Materiałs (R)	MAT
3	Stock (A)	SKL
4	Klients (R)	KLI
5	Order (A)	ZMW
6	Employes (R)	PRC
7	Departments (R)	DZL
8	Workplace (R)	STN
9	Mashins (R)	HRD
10	Programmes (R)	SFT
11	Computer Software A)	HIS
12	Technological Itinerary (A)	WTW

Problem to solve a specialist in the domain area

The selected company is an advertising agency dealing with IP (e-books, multimedia presentations, websites, etc.). The IS is designed to help manage (including control and supervise) workflows, and it also collects information on procurement and customers, product development, and tools used in product development agencies. Main offers of printing services:

    a. Design

    b. Digital processing, scanning

    c. Computer composition

    d. Exposure, printing (friendly printers)

    e. Bookbinding

Sample products: flyers, posters, calendars, notebooks, catalogues, books, magazines, etc. (formats and types: optional).

Let's examine the graph of relationships for modified example 1, represented in Fig. 4.33.

In Table 4.8 a set of IOs (real (R)) and abstract (A))is represented.

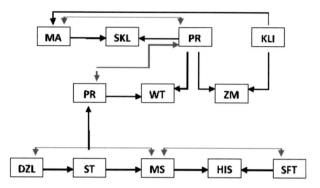

Figure 4.33 Graph of relationships for simplified example 1 (with $M : M$ relationship).

Problem to solve a system analyst

1. *Structure of individual attributes:*
   - Product code (4 digits, the first two of which indicate the code of the department that produced the product, and the other two the type of product that was produced)
   - Material code (6 digits, the first two indicate the type of material, and the remaining 4 are the customer's code from which the material comes)
   - Client code (a 4-digit code assigned to each new client of the advertising agency)
   - Order code (6 digits, the first 4 is the customer's ordering code, and the next 2 is the customer's order number)
   - Other codes
2. *Identification of types of real relations between IOs* (Table 4.9)
3. *Defining the relationship*
   - Products (PRD) – Materials (MAT) – $M : M$ relation (many products may consist of many materials, the Composition object (SKL) is used to solve this relationship on $1 : M$ and $M : 1$ relationships).
   - Products (PRD) – Composition (SKL) – $1 : M$ ratio (one product may be subject to multiple compositions, the maximum number of compositions is as many as the materials the customer will provide to make a given product).

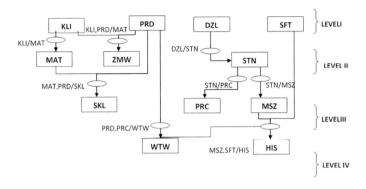

Figure 4.34 The canonical structure of relational data model.

- Products (PRD) – Customers (KLI) - $M : M$ relationship (many products can be associated with many customers, to solve this relationship on the $1 : M$ and $M : 1$ relationship, the Orders object (ZMW) is used). Let's examine the graph of relationships, for example.
- Others

4. *The canonical structure of the RDM*
   A graph of relationships for example is represented in Fig. 4.33. The canonical structure of the RDM is represented in Fig. 4.34. As we can see from the canonical structure of the RDM, we have created a four-level structure. At its top level are the following objects: *Customers*, *Products*, *Divisions* and *Software*. The second level are the objects *Materials*, *Orders*, and *Posts*. The third level are the objects *Composition*, *Workers* and *Machines*. At the last level are two objects, *Creation* and *Software*.

4.4.9 Typical Queries and Needed for Their Implementation of Functional Links

The following are examples of queries that can be directed to the IS you are creating. They are represented by the dependencies between objects and in the form of SQL queries.

Table 4.9 Matrix of real relations between information objects

	PRD	MAT	SKL	KLI	ZMW	PRC	DZL	STN	MSZ	SFT	HIS	WTW
PRD	–	M : M	1 : M	M : M	1 : M	M : M						1 : M
MAT	M : M	–	1 : M	M : 1								
SKL	M : 1	M : 1	–									
KLI	M : M	1 : M		–	1 : M							
ZMW	M : 1			M : 1	–							
PRC	M : M					–		M : 1				1 : M
DZL							–	1 : M				
STN						1 : M	M : 1	–	1 : M			
MSZ									–	M : M	1 : M	
SFT									M : M	–	1 : M	
HIS									M : 1	M : 1	–	
WTW	M : 1					M : 1						–

a) Find information about all machines in the publishing house at the workstation (assuming that the post is assigned *Job Code = 1*).

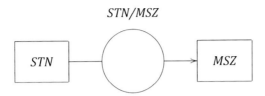

STN/MSZ

SELECT machines [machine codes], machines [machine name], machines [machine type] *FROM* machines *WHERE MACHINES* [Position Code] = 1;

b) View information about all machines used in the publishing house, including software installed on them.

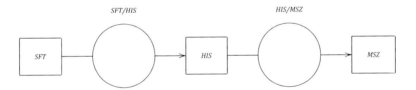

SFT/HIS   *HIS/MSZ*

SELECT machines [machine id], machines [machine name], machines [machine type], programs. [program name]*FROM* ((*INNER JOIN* Software ON [Machine Code]) *INNER JOIN* Programs *ON* [Program Code] *GROUP BY MACHINES* [machine code];

c) View all employees belonging to a department (e.g. *DTP, dtp code is* 2).
*SELECT * FROM* Employees *WHERE* [Division Code] $= 2$

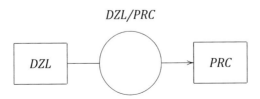

DZL/PRC

d) Display the order history of the selected customer.

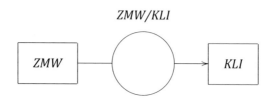

ZMW/KLI

SELECT * *FROM* (Customer *INNER* Order)
WHERE Clients. Name = 'Kowalski';

e) Indicate the material from which the product originated and the employees involved in its creation (suppose it is a *product of code* 3).

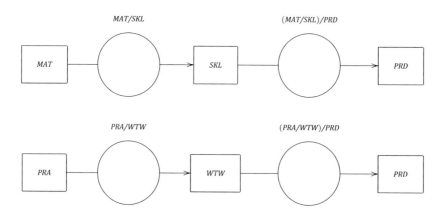

SELECT Material. [Material Name], Employees.Im, Employees.Name
FROM ((*INNER JOIN* Materials *ON*)) *INNER JOIN* Products *ON* [Product Code] *INNER JOIN* Production *ON* [Product Code]) *INNER JOIN* Employees *ON* [Employee Code] *WHERE* [Product Code] = 3;

f) Print a series of technological operations (TOs) for the production of the product (code 2), together with the names of the employees performing the individual operations.

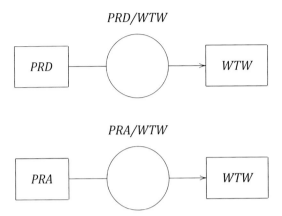

SELECT Manufacturing [Number of technological operations], Manufacturing. [Name of technological operation], Employees.Name *FROM* (Employee *INNER JOIN* Products *ON* [Product Code]) *ON* [Employee Code] *WHERE* [Product code] = 2 *ORDER BY* Production. [Number of technological operation];

4.4.10 Conclusion

A reference model was created for the enterprise, which is a printing agency. The first stage was to determine both real and abstract objects (created mainly to eliminate possible many-to-many relationships). Twelve basic objects (8 real and 4 abstract) have been determined in the model.

The next stage was to present relationships between the determined objects, as you can see even such a small enterprise as a printing agency can contain complex relationships. They were presented in the form of a relationship matrix and in the form of a relationship graph.

The development of the canonical form of the domain area ontological model, including the use of an appropriate algorithm, led to the creation of a hierarchical form of the model. A hierarchical four-level graph of real relations between objects was created, which shows exactly how the relationships between objects look like.

After identifying the relationships between the objects, the next task to complete was to propose typical queries. In an ontological model of RDB there are six such queries that can illustrate the scope of the designed IS. Thanks to these queries, the system user can find out what data he/she can extract from the system and what processes can be controlled through it.

The designed IS for a printing agency is designed to improve the flow of information and works. In addition, it shows the structure of the agency and helps to supervise the work of the company. It is definitely a system that will improve management of the organisation.

4.5 Case Study 2 Ontological Model of the Object-Oriented Database to Assess Acquired Professional Competence

4.5.1 Problem of Compatibility of Competencies in Professional Learning

4.5.1.1 Introduction

In the life of every person come certain key moments which often include the need for a decision to be taken. One such moment is the day we decide to dedicate the next few years of our lives to studying. This day may as well influence the rest of our lives. Choosing the right direction, the right university, the right education offer, gives us a better starting position in the job market afterwards. The main problem that occurs here is how to be sure we made the right choice.

The situation we discussed is somewhat complicated – the range of alternatives and factors that should be taken into account when taking a decision is great. There is a situation that requires IS solutions and gives the opportunity to automate the process to analyse the adjustment of educational offers to the requirements of the labour market in terms of acquired competencies and technologies, as well as the individual preferences of the decision maker (Fig. 4.35).

It was established that the gathering of information on competencies is best served with an ontological model in the form of

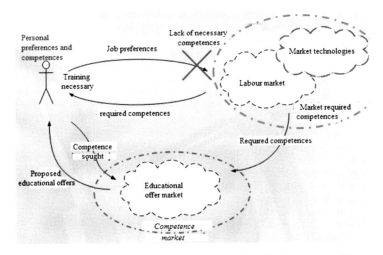

Figure 4.35 Place of an IS supporting the choice of educational offers.

an object database developed for this purpose, which gives more possibilities than simply comparing two sets of competencies.

Due to the introduction of the *Bologna Process* ("Bologna Working Group on Qualifications Frameworks", 2005), one can study in any area within united Europe, with a guarantee that the obtained results will be recognised in all other countries that accept the Bologna regulations. This means that not only possibilities grow in numbers, but also competition between education institutions increases, which allows us to assume that so will the quality of education. Nevertheless, there is still no easy way to analyse the available data and choose the option that gives the highest probability of actually getting a job after graduation. The fact of having a diploma becomes less important than the information about what this diploma represents and thus what competencies one has obtained during the education period.

The need to have a uniform look at education through the point of view of actual achievements and competencies leads to the development of several different standards and frameworks, such as the Qualifications Framework of the *European Higher Education Area* (*EHEA*) ("Bologna Working Group on Qualifications Frameworks", 2005) and resulting national qualification frameworks and standards, as developed, for example, by the Polish Ministry of Science and Higher Education (MNiSW, 2007), or resulting European standards for different education fields, as developed within the Tuning Project (Tuning, 2007). Accordingly, standard descriptions were created to describe the required competencies, abilities and knowledge connected to different jobs and positions (e.g. as developed by the Polish Ministry of Labour and Social Policy (MRPiPS, 2016) or within projects like European ICT Jobs (CompTrain,2008).

Not only the content-related meaning of competencies connected to a job or a field of study is being standardised, the same thing is happening regarding the way of describing competencies, requiring a comparison between them. This includes such standards and approaches as the reusable competency definition (RCD) ("IEEE LTSC", 2007), HR-XML ("HR-XML Consortium", 2006), OntoProPer (Sure *et al.*, 2000; "Advanced Distributed Learning Initiative", 2001), as well as the work of the TENCompetence project ("IEEE LTSC", 2006) and related works (e.g. De Coi *et al.* (2007).

The existence of all the mentioned standards and frameworks creates a basis for reference and comparison. However, there is still no mechanism allowing for automation of this process. Considering the dynamic changes of the requirements of the job market (which respond pretty quickly to new developments, new methodologies, standards or technologies) and the much slower response of the education market (due to more restrictions connected to education plans, legislation, etc.), one has to make sure that the education he/she decides to complete will be as close to the situation on the job market after graduation as possible. And the best way to do this would be to use the main tool present on both markets right now – the competencies – and to analyse their compatibility in an efficient, automatic way, which includes also the factor of presence, or lack thereof, of new developments in the considered domain.

4.5.1.2 Presentation of the problem

There are several situations in which we can see the use for a mechanism that allows assessing the *compatibility of competencies*, . Let us consider them from the point of view of who they relate to:

- University candidate – evaluating the compatibility of an education offer with the desired job profile

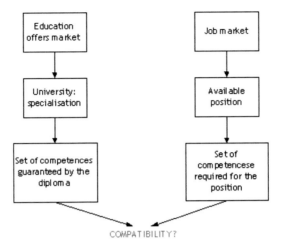

Figure 4.36 The problem of competence compatibility.

- Graduate – continuing education (lifelong learning), searching for employment
- University – assuring a competitive position among other universities
- Employee – choosing a vocational development path, changing jobs
- Enterprise – creating a staff development plan, looking for an employee with the desired profile

All of these cases include one additional desired aspect of *assessing the compatibility of competencies* – the possibility to assess the topicality of the considered competencies regarding a chosen point of reference. This aspect is especially important when beginning higher education, as it helps in assessing usability of the knowledge and abilities that are to be acquired in the education process once this process is completed. High usability of the acquired knowledge is undoubtedly good motivation for starting education, and a good justification for all the related cost.

From this point of view we can say that what has to be compared are competencies represented (guaranteed) by a graduation diploma and competencies required on the market to fulfil a certain job, as illustrated in Fig. 4.36.

Defining compatibility allows for creating a sort of bridge between both sets of competencies, so that they can be mutually understood and related to.

4.5.1.3 Proposition of a comparison model

In order to solve the stated problem the following have to be considered:

- Current standards regarding competencies and vocational qualifications (mentioned in Chapter 2,
- Contexts of both sets of competencies must be similar (otherwise a layer of common understanding cannot be created).

Additionally, to evaluate whether a certain set of competencies is up to date, information about the technologies and methodologies involved in it, regarding the current trends, should be included.

Currently, the comparison of competencies and evaluation of their compatibility is performed by people, in one-time actions (as needed), often giving different results each time and requiring a significant constraint on the possible number of options taken into account in one analysis. Therefore, it is important to create an automated comparison mechanism.

Competencies are usually reflected in textual form; thus it is important to establish their semantic meaning. The only instrument that allows defining borders of semantics is ontology (Gómez-Pérez *et al.*, 2004). Competencies can be compatible only if they belong to the same ontology.

Since in order to allow an automatic comparison, competencies have to be stored and described in some form that facilitates that, after considerations and studies, the authors decided to use a dedicated object-oriented database. As a static object model of the current state of competencies, the object database allowing for comparison of two competence sets has the following characteristics:

- provides the possibility to describe and store data structures, as well as methods and procedures for their usage within one object,

- gives the possibility to inherit characteristics of individual objects due to a description of their interrelations (e.g. an education offer can only describe its content, without repeating all information about the university it is connected to),
- is flexible; therefore when the need arises, the existing structure and hierarchy can be adjusted and changed, elements can be added, deleted or modified.

Additionally, the collected data can be used also for a retrospective analysis of the situation in the education and job market and its changes over time.

4.5.2 Ontological Model of Object-Oriented Database

Object-oriented database to assess acquired professional competence:

- Allows selection of a set of competencies for individual user's needs – for example, in the case of ODL the student has a big impact on determining the curriculum.
- Allows you to assess the perspectives of a certain set of competencies – determine its topicality on the foreseen time horizon.
- Ordered collected information gives the possibility of retrospective analysis, i.e. creating various statistics.

An object-oriented database represents an ontological model of the current state of the competence space and allows the description of the process of comparing two sets of competencies.
It has the following features:

- Provides the ability to describe and store data structures and methods/procedures for their use in a single object
- Gives the possibility to inherit the features of individual objects thanks to the description of their relationships
- Is flexible, so you can change the existing structure and hierarchy, introduce new elements or delete/modify existing ones

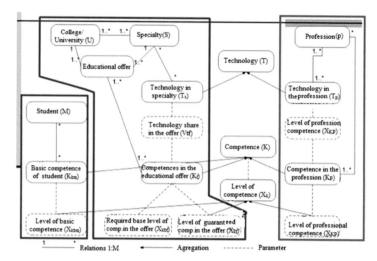

Figure 4.37 Structure of an object-oriented database.

The *ontological model of database* includes information about the labour market (*red border*), the education market (*blue*) and the university candidate (*violet*), as shown in Fig. 4.37. With this assumption, it is possible to develop procedures for evaluating the educational offer.

The figure shows the class diagram in UML notation, representing the proposed structure of the object-oriented database corresponding to the previously described ontological model.

This database can also be understood as a base of facts regarding the considered area of competencies, together with procedures used for their processing and analysis.

The considered database should include information describing the presented market situation and allowing for finding solutions of the described problems regarding the compatibility of competencies. Additionally, to evaluate whether a certain set of competencies is up to date, information about the technologies and methodologies involved in it, regarding the current trends, should be included.

The proposed content structure of the database was presented in Fig. 4.38.

Using the language of mathematics to described the proposed object model, we received the following formal model:

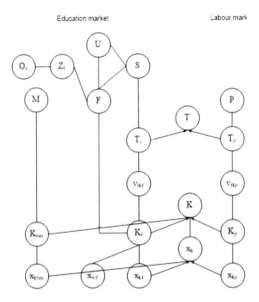

Figure 4.38 Ontological model of an object-oriented database.

1 *General assumptions*

   $K = \{k\}, k = 1, …, k^*$ – competence set

   All competencies belong to one common set of competencies. Depending on whether we look at them from
   the point of view of the job market or of the education
   market, we will talk about, respectively, the required and
   the guaranteed competencies. A given competence may
   be regarded differently on each market (e.g. fewer or
   different technological references), or under a different
   name, but still remain the same competence.

   Example: 'knowledge of English language'

   x_k – strength of competence k

   As the strength of a competence we understand the level
   of fluency in the competence (see e.g. in (Wang and Wang,
   1998) or (Yu and Zhang, 1990)) described by projecting
   a linguistic or numerical description of the evaluation
   of this level on a unified scale. In order to guarantee
   accuracy, a detailed method of transforming other scales
   into the uniform one should be developed with the help

of specialists of the considered domains, cognitive science, semantics, etc.

Scope :< 0...1 >

0 signifies lack of any knowledge regarding the competence, while 1 signifies a complete set of knowledge and abilities included in the competence (expert level).

Example: 'fluent knowledge of English' can be described, for example, as $k =$ 'knowledge of English' and $x_k = 0.9$ (level similar to a native speaker).

$T = \{t\}, t = 1, ..., t^*$ – set of technologies, methodologies and tools connected to the competence.

All technologies, methodologies and tools belong to one common set. Their connection to competencies, however, maybe different depending on where they are used.

We talk about technologies, methodologies and tools since they can all become outdated; thus their connection to required and guaranteed competencies is important for evaluating the usability of an education offer.

Example: a project manager has to know such project methodologies, for example the *ProjectManagementBodyofKnowledge* (PMBOK), for which every few years an update is released; therefore knowledge about the version might be quite important.

2 *Job market $P = \{p\}, p = 1, ..., p^*$ – set of professions*

As a profession we understand a job/position sought on the market.

$K_p = \{k_p\}, k_p = 1, ..., k_p^*, K^p \subset K$ – set of competencies k required in profession p

Each profession is connected to having a certain set of competencies, which is usually described in a linguistic way, e.g. 'very good knowledge of English'.

x_{kp} – strength of competence k required in profession p

For each competence required in a certain profession, a minimum acceptable strength of the competence should be specified. Scope: $(0, 1)$ – competencies with strength 0 are simply not considered.

$Tp = \{t_p\}$, $t_p = 1,..., tp^*$, $Tp \subset T$ – set of technologies, method-ologies and tools connected to profession p

> Each profession is connected to knowledge about a certain set of technologies, methodologies and tools required and the ability to use them. For, example, for a computer graphic designer, this might be Photoshop or Corel, while for a project manager it could be PMBOK or PRINCE2.

v_{tkp}, $t_{kp} = 1,..., t_{kp}^*$; $k_p = 1,..., kp^*$; $p = 1,..., p^*$ – participation of technology t in the required competence k_p

> Scope: $(0..1)$
> 0 means that a certain technology (methodology, tool) is not used in this competence; thus we do not consider this option; 1 signifies that the profession is dominated completely by one technology.

3 *Education market $S = \{s\}$, $s = 1,..., s^*$* – set of specialisations
$U_s = \{u\}$, $u_s = 1,..., u_s^*$ – set of universities which offer speciali-sation s
f_{us}, $u = 1,..., u^*$, $s = 1, ...s^*$ – education offer of university u referring to specialisation s
$Z_f = \{z_f\}$, $z_f = 1,..., z_f^*$ – set of subjects connected to offer f_{us} (curriculum)

> Information about the set of subjects can help in defining the set of competencies consisting for the considered specialisations.

Q_z – parameters of subject z in offer f_{us} (e.g. the number of hours, form of classes, form of evaluation, etc.)

> Parameters of a subject can be used for establishing the strength of a competence guaranteed by an education offer.

$K_f = \{k_f\}$, $k_f = 1,..., kf^*$, $K_f \subset K$ – set of competencies guaran-teed by offer f_{us} when teaching speciality s at university u.
x_{kf} – strength of competence k guaranteed by offer f_{us} after graduation

> Each offer should define the level of mastery of a competence upon graduation. One can assume that the simple fact of teaching a certain scope of knowledge guarantees its mastery at a basic level. The strength of

the competence may be established on the basis of the number of hours of teaching a given subject. *Scope*: $(0, 1)$ – competencies with strength 0 are simply not considered.

x_w – strength of competence k required in order for a candidate to be accepted for studying specialisation s in the frames of the education offer f_{us}

Each education offer can be characterised by a set of requirements, without the fulfilment of which it is impossible to be accepted. The requirements refer to the scope and the level of competencies one needs to possess before starting the studies. The regarded set of competencies is K_f.

Scope: $(0, 1)$ – competencies that are not important in the recruitment process have the required level of competence strength 0.

$T_s = \{ts\}, t_s = 1,..., t_s^*, T_s \subset T$ – set of technologies, methodologies and tools connected to specialisation s

A certain specialisation may be connected to different technologies, but their participation in this speciality depends on a specific education offer.

$v_{tkf}, t = 1,..., ts^*; \ k = 1,..., kf^* \ ; \ f = 1,...,f_{us}^*$ – participation of technology t_s in competence k_f guaranteed by offer f_{us} when studying specialisation s at university u

Scope: $(0..1)$

0 means technology t_s is not present in offer f_{us}, but the technology might be present in a different offer, so it is still considered, while 1 means the offer is completely dominated by the technology.

4 *Candidate for studies*

$M = \{m\}$ – set of candidates for studies

Although in ODL we talk rather about student flow than about a set of students, in the situation for which the described model was developed, at a single time we are only dealing with one candidate for studies. Using the concept of a set has meaning mainly in case of creating statistics based on the data collected in the database, while the flow of students itself is never considered.

Each candidate is characterised by:

$K_{bm} = \{k_{bm}\}$, $k_{bm} = 1, \ldots k_{bm}^*$ – set of basic competencies of candidate m

> At any given time each person possesses a certain set of competencies on the basis of which his/her knowledge and abilities can be further developed. We call these the basic competencies.

x_{kbm} – strength of the basic competence k_b of candidate m

> Each of the possessed competencies was mastered to some level. This level of mastery is reflected by the strength of the basic competence.
>
> *Scope: (0,1)* – competencies with strength 0 are simply not considered.

5 *Procedures*

Certain procedures can be identified for performing the most important tasks on the data included in the database. These include:

a. *Coverage of the minimum required for being accepted for studies by the basic competencies of a candidate*

Application:

> This procedure is used to assess whether a candidate can start studies proposed by a given education offer.

Input data:

> m – candidate for studies
>
> u – specific university
>
> s – specific specialisation
>
> f_{us} – specific education offer for specialisation s at university u
>
> K_f – set of competencies guaranteed by offer f_{us}
>
> x_w – strength of competence k_f required to be accepted for studies
>
> K_{bm} – set of basic competencies of student m
>
> x_{kbm} – strength of competence k belonging to set K_{bm}

Procedure:

> $\forall k_f, \in K_f \exists k_{bm} \in K_{bm}$, such as $k_f = k_{bm}$ and $x_{kbm} \geq x_w$

Result:

The result is given as **true/false**, showing whether a candidate possesses the required set of competencies at the required level.

b. *Concordance of an education offer with the requirements of the chosen profession*

Application:

The procedure is used for evaluating the level of concordance of competencies guaranteed by the education offer with the competencies required for a certain profession. This refers to the competencies themselves as well as their strength and the participation of technologies, etc., in them.

Input data:

p – specific profession

u – specific university

s – specific specialisation

f_{us} – specific education offer for teaching specialisation s at university u

K_f – set of competencies guaranteed upon graduation from studies described by offer f_{us}

x_{kf} – strength of competence k belonging to set K_f

K_p – competencies required for profession p

x_{kp} – strength of competence k belonging to set K_p

v_{tkp} – participation of technology t in competence k connected to profession p

v_{tkf} – participation of technology t in competence k included in offer f_{us}

Procedure:

$$FOR(i = 0...k_p^* \ AND \ j = 0...k_p^*; \ i++j++)$$
$$IF \ \exists(k_i \in K_f \ AND \ k_j \in K_p$$
$$such \ that \ k_i = k_j) \ THEN$$
$$IF \ (x_{ki} \geq x_{kj}) \ THEN$$
$$IF \ \forall z = 1...t_p*, \ v_{zki} \geq z_{zkj}) \ THEN$$
$$NoCoveredCompetencies ++;$$
$$ELSE$$

$$NoComWithLowerTechPart\ ++;$$
$$ELSE$$
$$NoComWithLowerStrength\ ++;$$
$$ELSE$$
$$NoMissingCom\ ++;$$
$$END\ FOR$$
$$FullyCovered =$$
$$(NoCoveredCcompetencies/k_p) * 100\%$$
$$NotCoveredTech =$$
$$(NoComWithLowerTechPart/k_p) * 100\%$$
$$NotCovStength =$$
$$(NoCompWithLowerStrength/k_p) * 100\%$$
$$NotCoveredAtAll =$$
$$(NoMissingCom/k_p) * 100\%$$

Result:

The result defines the percentage of competencies required in the profession that are fully covered by the guaranteed competencies and the percentages of competencies for which the participation of technologies or the strength level are too small, as well as the percentage of competencies not covered at all.

It is possible to further split the calculations in order to identify the participation of technologies not covered by the education process in a level required for the considered profession.

Several other interesting procedures can be created for the described object model:

Application:

The procedure may be used for evaluating the scope of knowledge increase as a result of competing the education as described in the education offer, by calculating the number of competencies not possessed by the candidate or not mastered to the offered level (optionally: including outdated content).

c. Coverage of a profession by a certain education offer (or specialisation, in which case a small change in the mathematical assumptions would be needed).

Application:

The procedure could be used for finding education offers (specialisations) that are, on the level of competencies, most consistent with a given profession.

d. *Defining the difference between competencies offered in the frames of different education offers.*
Application:

The procedure could be used by universities to evaluate difference with other offers in order to become more competitive in a certain field or to identify a 'niche' which could be filled in by a new offer.

4.5.3 Conclusions

As was discussed in the chapter, providing an automated tool for evaluating the compatibility of competencies is the first step in facilitating the choice of higher education. The proposed model solves the basic issue of not only assessing whether a certain set of guaranteed competencies contains competencies required on the job market, but also by including the technological aspect helping decide whether the education offer is up to date concerning the content of each competence. The next step would be incorporating the model in a system that would also give each user the possibility to personalise the results.

4.6 Case Study 3: Ontological Model of Supply Chain Management

'Although human genius through various inventions, makes instruments corresponding the same ends, it will never discover an invention more beautiful, nor more economical than does nature, because in her inventions nothing is lacking and nothing is superfluous'

Leonardo da Vinci, 16th century

4.6.1 What Is Supply Chain Management?

4.6.1.1 Environmental issues of manufacturing

A supply chain (SC) is a system of organisations, people, technology, activities, information and resources involved in moving a product or service from supplier to customer. SC activities transform natural resources, raw materials and components into a finished product that is delivered to the end customer (Fig. 4.39).

A typical SC begins with ecological and biological regulation of natural resources, followed by the human extraction of raw material, and includes several production links (e.g. component construction, assembly and merging), moves on to storage facilities and finally reaches the consumer.

An extended enterprise is a loosely coupled, self-organising network of businesses that cooperate to provide product and service offerings. Therefore the exchanges encountered in the SC will therefore be between different companies that will seek to maximise their revenue within their sphere of interest.

4.6.1.2 Environmental issues of manufacturing

ISO 14001 Environmental Management Systems supposes environment issues of manufacturing, depicted in Fig. 4.40.

Participants of SC: customer – an integral part of SC, manufacturers, suppliers, transporters, warehouses, retailers, wholesalers/distributors

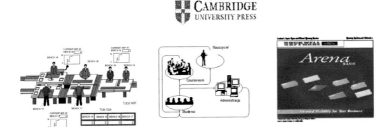

Figure 4.39 Cambridge University Press.

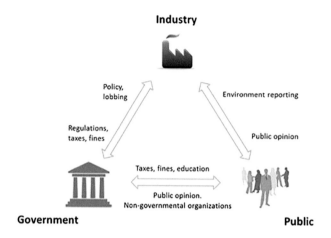

Figure 4.40 Environmental issues of manufacturing.

Functions of SC: receiving and filling of customer requests, new product development, marketing, production operation, distribution, finance and customer services

Each stage in an SC is connected through the flows of materials, products, information and funds. These flows often occur in both directions.

4.6.1.3 Goal of supply chain

The primary purpose of an SC is to maximise the overall value generated. The value an SC generates is the difference between what the final product is worth to the customer and the costs the SC incurs in filling the customer's request. It is known also as the SC surplus or SC profitability.

SC surplus can be defined as the difference between the revenue generated from the customer and the overall cost across the SC.

SC decisions have a large impact on the success or failure of each firm because they significantly influence both the revenue generated and the cost incurred.

4.6.1.4 Supply chain management

Supply chain management (SCM) is the management of a network of interconnected businesses involved in the ultimate provision of

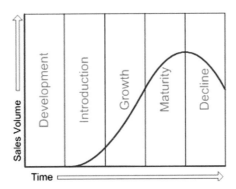

Figure 4.41 Product life cycle.

product and service packages required by end customers. Supply Chain Management spans all movement and storage of *raw materials, work-in-process inventory and finished goods* from the point of origin to the point of consumption.

Another definition is provided by the *APICSDictionary* when it defines SCM as the *'design, planning, execution, control, and monitoring* of SC activities with the objective of creating net value, building *a competitive infrastructure, worldwide logistics*, synchronising supply with demand, and measuring performance globally.'

SCM has to manage *the flows of product, information and funds* to provide availability to the customer while keeping costs low.

Three key SC decisions phases Successful SCM requires many decisions relating to the flow of *information, products and funds.*

Each decision should be taken to raise the SC surplus. These decisions fall into categories or phases, depending on the frequency of each decision and the time frame during which a decision phase has an impact. As a result, each category of decisions must consider uncertainty over *the decision horizon.*

SC decision phases may he categorised as *strategy*, *planning and operational* ones, depending on the time frame during which the decisions taken apply.

a) *SC strategy or design*

The strategic decisions are typically made for the long term (a matter of years) and are very expensive. They are the following:

- What the SC's configuration will be
- How resources will be allocated
- What processes each stage will perform
- How to structure the SC

b) *SC planning*

The time frame considered is *a quarter to a year. The goal* of planning is to maximise the *SC surplus* that can be generated over the planning horizon, given the constraints established during the strategic phase. *As the result* of the planning phase, companies define a set of operating policies that govern short-time operations.

c) *SC operation*

The time horizon here is weekly or daily, and during this phase companies take decisions regarding individual customer orders.

The goals of SC operations are:

- To handle incoming customer orders
- Allocate production resources (PR) to individual orders
- Set a date that an order is to be filled
- Set the mode and schedule of delivering

Life cycle of a supply chain   A cycle view of an SC divides processes into cycles, each performed at the interface between *two successive stages* of an SC (Fig. 4.41). Each cycle starts with an order placed by one stage of the SC and ends when the order is received from the supplier stage. A push/pull view of an SC characterises processes based on their timing relative to that of a customer order. *Pull processes* are performed in response to a customer order, whereas *push processes* are performed in anticipation of customer orders.

4.6.1.5 Supply chain classification

All SC processes can be classified into *three macro-processes*, depending on whether they are (1) at the customer or (2) supplier interface or are (3) internal to the company.

The customer relationship management (CRM) macro-process consists of all processes at the interface between the company and the customer that work *to generate, receive and track* customer orders.

The internal supply chain management (ISCM) macro-process consists of all SC processes that are internal to the company and *work to plan for and fulfil* customer orders.

The supplier relationship management (SRM) macro-process consists of all SC processes at the interface between the company and its suppliers that work *to evaluate and select suppliers* and then *source goods and services from them.*

What about logistics?

Logistics is the management of *the flow of goods, information and other resources, including energy and people, between the point of origin and the point of consumption* in order to meet the requirements of consumers. Logistics is *a channel of the SC* which adds the value of time and place utility. Today the complexity of production logistics (PL) can be modeled, analysed, visualised and optimised by plant simulation software.

Logistics management is that part of the SC which plans, implements and controls the efficient, effective forward and reverse *flow and storage of goods, services and related information* between the point of origin and the point of consumption in order to meet customer and legal requirements. A professional working in the field of logistics management is called a *logistician.*

What's the difference? There is often confusion over the terms 'supply chain' and 'logistics'. It is now generally accepted that the term 'logistics' applies to activities within one company/organisation involving distribution of product, whereas the term 'supply chain' also encompasses manufacturing and procurement and therefore has a much broader focus as it involves multiple enterprises, including suppliers, manufacturers and retailers, working together to meet a customer need for a product or service.

The term 'Production logistics' is used for describing logistic processes within an industry. *The purpose of PL* is to ensure that each machine and workstation is being fed with the right product in the right quantity and quality at the right point in time.

The problem of PL is not the transportation itself, but to streamline and control the flow through the value adding processes and eliminate non-value-adding ones.

PL can be applied in existing as well as new plants. Manufacturing in an existing plant is a constantly changing process. Machines are exchanged and new ones added, which gives the opportunity to improve the PL system accordingly. Production logistics provides the means to achieve customer response and capital efficiency.

PL is getting more and more important with decreasing batch sizes. In many industries (e.g. mobile phones) batch size is a short-term aim. This way even a single customer demand can be fulfilled in an efficient way. Track and tracing, which is an essential part of PL, due to product safety and product reliability issues, is also gaining importance, especially in the automotive and medical industries.

What does SCM mean with regard to manufactured goods? What about services?

The primary objective of SCM is to fulfil customer demands through the most efficient use of resources, including distribution capacity, inventory and labour. In theory, a SC seeks to match demand with supply and do so with the minimal inventory.

4.6.1.6 Supply chain optimisation

Optimisation of the SC includes:

- Liaising with suppliers to eliminate bottlenecks
- Sourcing strategically to strike a balance between lowest material cost and transportation
- Implementing just-in-time (JIT) techniques to optimise manu-facturing flow
- Maintaining the right mix and location of factories and warehouses

Optimisation serves customer markets using location/allocation, vehicle routing analysis, dynamic programming and, of course, traditional logistics optimisation to maximise the efficiency of the distribution side.

Supply chain management problems Supply chain management must address the following problems:

Distribution network configuration:

- Number, location and network missions of suppliers
- Production facilities, distribution centres, warehouses, cross-docks and customers

Distribution strategy:

- Questions of operating control (centralised, decentralised or shared)
- Delivery scheme, e.g. direct shipment, pool point shipping
- Mode of transportation, e.g. motor carrier, including truckload, less than truckload (LTL), parcel, railroad
- Intermodal transport, including trailer on flatcar (TOFC) and container on flatcar (COFC) (container on flatcar)
- Ocean freight, airfreight, replenishment strategy (e.g. pull, push or hybrid)
- Transportation control (e.g. owner-operated, private carrier, common carrier, contract carrier)

Trade-offs in logistical activities: The aforementioned activities must be well coordinated in order to achieve the lowest total logistics cost. Trade-offs may increase the total cost if only one of the activities is optimised. For example, full truckload (FTL) rates are more economical on a cost per pallet basis than LTL shipments. If, however, an FTL of a product is ordered to reduce transportation costs, there will be an increase in inventory-holding costs, which may increase total logistics costs. It is therefore imperative to take a systems approach when planning logistical activities. These trade-offs are key to developing the most efficient and effective Logistics and SCM strategy.

Information: Integration of processes through the SC to share valuable information, including demand signals, forecasts, inventory, transportation, potential collaboration, etc.

Inventory management: Quantity and location of inventory, including raw materials, work-in-progress (WIP) and finished goods.

Cash-flow: Arranging the payment terms and methodologies for exchanging funds across entities within the SC.

Supply chain execution: Managing and coordinating the movement of materials, information and funds across the SC. The flow is bi-directional.

Why has there been so much interest in SCM over the past few years?

There are three factors (or problems) which make the idea of SCM very actual:

1. *Problem of production process (PP) integration*
   It was observed that automation on the level of separate stages of the entire PP – mainly supplying, manufacturing, delivering – is not effective from the point of view of the production cost, timeliness and reliability and often leads to customers loss. Searching for a reduction of the production cost requires a more wide-scale context of the automation problem, for example analysis of material and financial flows between different participants of the integrated PP.

2. *Problem of PP optimisation*
   Very often optimisation of separate sectors of the PP leads to contradictory goals and negative results. It is not possible to optimise or improve only some parts of the PP. Designers of new products, new technologies and new materials need an integrated approach, from supplying of raw materials to delivering the finished product to the consumer.

3. *Problem of widespread utilisation of IT*
   We can, in fact, say that IT has launched a new breed of SCM applications. The Internet and other networking links learn from the performance in the past and observe historical trends in order to identify how much product should be made, along with the best and cost-effective methods for warehousing it or shipping it to the retailer.

4.6.2 Mathematical Model of an Enterprise

The principal purpose of a *general mathematical model* (GMM) of intangible production (IP) is to describe and to offer a universal mathematical tool to the designer and manager of IP for decision-making. These decisions may refer to the topology and structure of IP as well as the physical and logical parameters of the medium for its realisation (hardware and software). GMM is aimed as the most effective organisation of IP.

A pre-requisite of universal model creation must be the classification of enterprises with intelligent production and the definition of parameters for each class. Nevertheless, independently of the enterprise class there exist three basic components of the general model of IP:

- Mathematical model of the production process (MMPP)
- Mathematical model of the enterprise environment (MMEE)
- Mathematical model of the distributed production network (MMDPM)

1. *MM of Production Process*

   The aim of the MMPP is to use of general principles of IP and private characteristics of an object to give a formalised description of a specific PP. The MMPP includes all parameters of production essential from a point of view of economic and organisational management by the enterprise.

   Now, let's submit the following definitions (Fig. 4.42):

   - Production process (PP)
   - Technological operation (TO)
   - Work articles (WA)
   - Production resources (PR)

   A PP is creating a completed product from materials, supplying items and semi-finished articles (called 'material resources') using certain PR. Besides a PP, there are other processes in enterprise, such as:

   - Logistics
   - Marketing
   - Acquisition and installation of equipment
   - Servicing and repair, etc

   A TO is an elementary (from the manager's point of view) PP that makes the transformation of some material resources to other ones. A full set of TOs can be divided into processing,

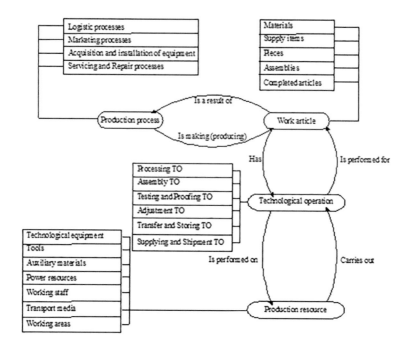

Figure 4.42 Semantic relationships (entity relationship diagram) of the production process.

assembly and control operations. Besides TOs, there are other kinds of operations at enterprises. These are the operations of:

- Proofing and testing
- Product transferring,
- Storage at central and local buffers
- Assembling and dispatching

Work articles are material resources that are completely used at the input of TOs or appear as a result of TOs. These are materials, supplying units, semi-finished articles, assemblies and completed articles.

Production resources are used to perform the TOs. These are technological equipment, tools, power, workers, stores and working areas.

A *structural graph of a product* (*SGP*) is a reflecting structure of some completed product. Vertices of the SGP are input items, and arcs are output items. Therefore, an SGP consists of constructive information about the structure of a product as well technological information about its manufacture.

Technological itinerary is a reflecting technology of the production of some piece (or semi-product), before the completed product. Vertices of the graph are TOs, arcs are sequential relations between operations. Therefore, the technological itinerary of piece consists of technological information about its manufacture.

A *TO* can be considered as an indivisible element of a PP, in other words, a brick, forming an MMPP. Each TO is characterised by the following parameters: (1) set of output items, (2) production resources, (3) rules of processing, (4) conditions of processing and(5) set of input items.

2. *Mathematical model of an enterprise environment*

The basic purpose of a *management information system* (MIS) for an enterprise is effective structure of production management with allowance of varying environmental conditions and shaping optimal performance evaluation, reaching the objectives of the enterprise in view.

The problem of management of an enterprise, realising the goods and services of mass demand, has a number of features:

    a) The large nomenclature of produced articles

    b) Ramified manufacturing structure of production and SC

    c) Dependence of production on the environmental conditions exhibited in effect of random factors

The scheme of the environment infrastructure of the enterprise is shown in Fig. 4.43.

4.6.3 Ontological Model of Distributed Supply Chain (DSC)

The ontological model of a distributed supply chain (in UML notation) is represented in Fig. 4.44. There are three kinds of relationships in an ontological model: generalisation, aggregation and structural relation 1:M (Fig. 4.45) A set of IOs in an ontological model is shown in Table 4.10. Relationships of information objects in the model are represented in Table 4.11.

Table 4.10 A set of information objects in an ontological model

N	Name of the object	Semantics of the object	Designation of the object	Minimal composition of attributes
1	O1 Market segment	A group of possible customers who are similar in their needs, age, education, etc.	MAS	*Type of production* Customer type The volume of the segment Segment dynamics
2	O2 Production type	—	PRT	*Production type ID* Right of arrival (accident, periodic, normal) Product character (deterministic, stochastic)
3	O3 Customer type	—	CUS	*Customer type ID* Customer characteristics (permanent, changeable) Customer category
4	O4 Order	—	ORD	*Order ID* *Date of order* Customer type Customer characteristics (Name, Address, Territories) Product type (Name, Product price, Due date) Production characteristics

(Continued)

Table 4.10 *(Continued)*

N	Name of the object	Semantics of the object	Designation of the object	Minimal composition of attributes
5	O5 Production chain	—	PRC	*Production network ID* The name of the production network The scale of the production network (LAN, MAN, WAN)
6	O6 Production network nodes	—	PRN	*Production network ID* *Node ID* The type of node (supplier, production centre, distributor)
7	O7 Production network channel	—	PRS	*Production network ID* Channel type (supplier - CP, CP - distributor)
8	O8 Supplier	—	SUP	*Node ID* *Supplier ID* Characteristics of the supplier (Name, Address, etc.)
9	O9 Production centre	—	PRC	*Node ID* *ID centres* Characteristics centres (Name, Location)

(Continued)

Table 4.10 *(Continued)*

N	Name of the object	Semantics of the object	Designation of the object	Minimal composition of attributes
10	010 Dystrybutor	—	DYS	*Node ID* *Distributor's ID* Characteristics of the distributor (Name, Location, etc.)
11	011 Delivery channel	—	DEL	*Supplier ID* *Channel ID* Channel capacity
12	012 Material delivery	—	DEM	*Supplier ID* *ID of the production centre* Delivery type Delivery date
13	013 Distribution channel	—	DEC	*Channel ID* *Supplier ID* Channel capacity
14	014 Product delivery date	—	PRD	*ID of the production centre* *Supplier ID* Product type delivery

(Continued)

Table 4.10 *(Continued)*

N	Name of the object	Semantics of the object	Designation of the object	Minimal composition of attributes
15	O15 Production resources of the company	—	PRR	*The type of resource* *(informative, technical, human, financial)* *Asset ID* The name of the resource Quantitative characteristics of the resource (Number or volume of resource, owner, etc.)
16	O16 Work	—	WOR	*Order ID* *Work ID* The name of the work Quantitative work characteristics (Number of pages, Format, etc.)
17	O17 Position	—	POS	*Production node ID* The name of the resource Quantitative characteristics of the resource (Number or volume of resource, owner, etc.)

Table 4.11 Relationships of information objects in the model

N	Object input	Object output	Type of Relations	Cardinal numer
1	Segment market	Type production	Generalisation	1-10
2	Segment market	Customer type	Generalisation	1-4
3	Type production	Order	Real relation	0-100
4	Customer type	Order	Real relation	0-10
5	Work	Order	Aggregation	0-10
6	Position	Working	Relationship real	1:M
7	Network nodes production	Product network	Relationship real	10-100 (depends on the scale)
8	Network channels production	The productive network	Generalisation	10-100 (depends on the scale)
9	Suppliers	Production Network Nodes	Generalisation	10-100 (depends on the scale)
10	Manufacturing centries	Production Network Nodes	Generalisation	1-10 (depends on the scale)
11	Distributors	Production Network Nodes	Generalisation	10-100 (depends on the scale)
12	Manufacturing centries	Positions	Relationship real	1:M
13	Manufacturing centries	Resources	Relationship real	10:100
14	Resources inf, equipment, human, financial	Resources	Agregation	10:1000
15	Delivery channels materials/parts	Channels	Generalisation	10-100 (depends on the scale)
16	Channels distribution	Channels	Generalisation	10-100 (depends on the scale)
17	Deliveries materials/parts	Channels supply	Relationship real	1:M
18	Deliveries products	Channels distribution	Relationship real	1:M

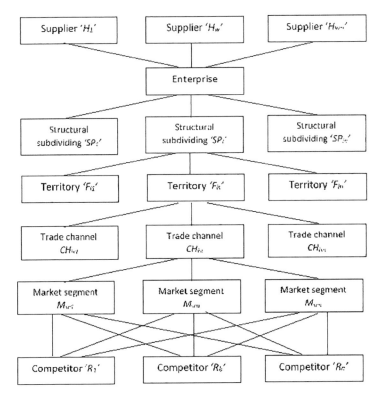

Figure 4.43 Scheme of the environment infrastructure of an enterprise.

4.6.4 Mathematical Procedures of a DSC Ontological Model

Database applications

1 *Parameters of the production network*
   $\{X, Y\}$ – graph, distorting the structure of the production network, where

   X is the set of graph vertices (06)

   Y is the graph edge set (07)

   $X = S \cup P \cup D$, the vertices of the graph, are the nodes of the production network, which can be of three types:

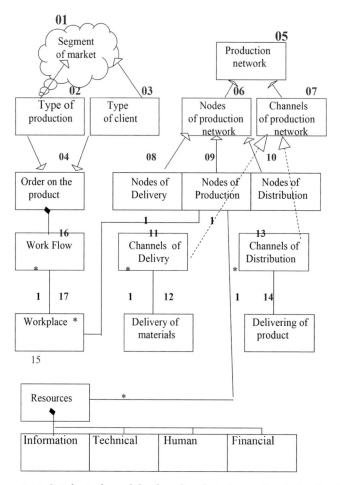

Figure 4.44 Ontological model of a distributed supply chain (in UML notation).

$S = \{s\}$ are the suppliers of materials and purchased parts ($O8$)

$P = \{p\}$ are the production centres (PC) ($O9$)

$D = \{d\}$ are the distributors ($O10$)

$Y = A \cup B$, the edges of the graph, are production network channels, which can be of two types:

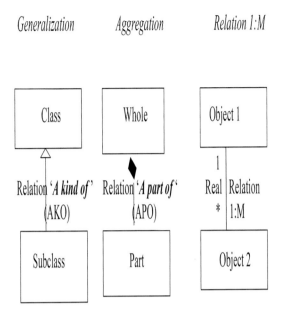

Figure 4.45 Three kinds of relationships in the ontological model.

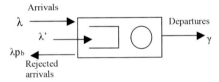

Figure 4.46 Structure of the production site $(M/M/1/S$ queuing system).

$A = \{a\}$ are material delivery channels from suppliers to CP ($O8$)

$B = \{b\}$ is the delivery of finished products from CP to distributors ($O10$)

$a = (s, p)$ is the delivery channel that connects the delivery node with the production node ($O8, O9$)

$b = (p, d)$ is the delivery channel that connects the production node with the distribution node ($O9, O10$)

$\mu(a), \mu(b)$ is the capacity of the delivery channel and distribution channel ($O11, O13$)

2 Vector of resources

$T_0 - 1$ is the planning period

$\overline{R} = (r_1, r_2, ..., r_k)$ is the vector of hardware and software resources, information and human and financial resources ($O15$)

$\overline{H}(p) = (h_1^p, h_2^p, ..., h_k^p)$ is the volume of resources allocated in the production node $p \in P$ for the planning period T_0 ($O15$)

3 *Order parameters*

$\Pi(n) = \{u(n), Q(n), T(n), V(n), U(n), \overline{Z}(n,r), \overline{W}(n)\}$ are order parameters, where

$u(n)$ is the type of the ordered product ($O4$)

$Q(n) = \{s(n), p(n), d(n)\}$ is the ordering scheme ($O4, O8, O9, O10$)

$s(n) \in S$ is the material/parts supplier's node

$p(n) \in P$ is the production node

$d(n) \in D$ is the product distribution node

$T(n) = (t_1^n, t_2^n)$ is the order execution interval ($O4$)

$t_1^n \in T_0$ is the date of arrival of the order

$t_2^n \in T_0$ is the finish time (delivery) of the order

$V(n)$ is the delivery volume ($O4, O8$)

$U(n)$ is the volume of product delivery ($O4, O10$)

$\overline{Z}(n) = (z_1^n, z_2^n, ..., Z_k^n)$ is the resource vector needed to complete the order ($O4, O15$)

$\overline{W}(n) = (w_1^n, w_2^n, ..., w_l^n)$ is the vector of resources needed to fulfil the order ($O16$)

$\tau(w_j^n), j = 1, 2, ..., l_n$ is the time of work ($O16$)

4 *Parameters of the production network operation*

a) Delivery time

Time of order fulfilment

$$\tau(n) = \tau^S(n) + \tau^P(n) + \tau^D(n),$$

where

$\tau^S(n)$ is the delivery time of materials/parts purchased for the channel a

$$\tau^S(n) = \frac{V(n)}{\mu(a)}$$

$\tau^P(n)$ is the product development time $u(n)$

$$\tau^P(n) = \sum_{j=1}^{l} \tau(w_j^n)$$

$\tau^D(n)$ is the time of delivering the product $u(n)$ to the customer after the channel b,

$$\tau^D(n) = \frac{U(n)}{\mu(b)}$$

b) Resources $r \in R$ needed in the production centre $p \in P$ for the planning period T_0

$$Z^r(p) = \sum_{n=1}^{N} \alpha_p^n z_r^n,$$

where

$n : (t_1^n, t_2^n) \in T_0$ is the collection of orders arriving for the period T_0

α_p^n is the symbol of Kroneker:

$$\alpha_p^n = \begin{cases} 1, & \text{if the order 'n'} \\ & \text{made in the production node 'p'} \\ 0, & \text{otherwise,} \end{cases}$$

z_r^n is the volume of the resource 'r' needed to perform the contract 'n'

c) Use of the resource $r \in R$ in the production centre $p \in P$ for the planning period T_0

$$\rho_r(p) = \frac{z^r(p)}{h^r(p)},$$

$z^r(p)$ is the demand in the resource 'r' in the production centre 'p'

$h^r(p)$ is the volume of the resource 'r' separated in the production centre 'p'

d) Bottleneck (BN) in the production network

$$p^{BN} = \max_{p \in P} \max_{r \in R} \rho^r(p),$$

where $\rho^r(p)$ is the use of the resource 'r' in the production centre 'p'

e) Probability of blocking the production node (Fig. 4.46)

$$Q_p^B = 1 - \frac{\mu_p(1 - Q_p^0)}{\lambda_p}$$

where

Q_p^B is the probability of blocking the node p

λ_p is the rate of arrival of client orders to the node p

μ_p is the rate of handling orders at the node p

Q_p^0 is the probability of an empty (free) state of the node p

f) $\tilde{\tau}^W$ average waiting time in the production node

$$\tau^W = \frac{N_q}{\lambda'} = \frac{\rho}{\mu - \lambda} - \frac{B\rho^{B+1}}{\mu - \lambda\rho^{B+1}}$$

4.7 Conclusion

Supply-chain management, techniques with the aim of coordinating all parts of an SC from supplying raw materials to delivering and/or resumption of products, tries to minimise total costs with respect to existing conflicts among the chain partners. An example of these conflicts is the inter-relation between the sales department desiring to have higher inventory levels to fulfil demands and the warehouse for which lower inventories are desired to reduce holding costs.

The integration of the different hierarchical decision-making levels involved within an SC is essential for its adequate management in dynamic and competitive markets. Any approach encompassing design issues, planning, coordination and responses to customer demands requires the consideration of huge amounts of data, which are a valuable source of information only if properly managed. But these data could also cause a lack of coordination if not stored and interpreted appropriately, so standardising information structures and tools to improve the availability and communication of data information between different hierarchical decision levels is essential. Thus, this work addresses the problem of making the best use of the ISs associated with an SC in order to improve the knowledge and information comprehension capabilities in the area of process systems engineering.

Chapter 5

Motivation Modelling in Open Distance Learning

Motivation aspects as interaction among teachers and students are essential for a successful learning process in an open distance learning (ODL) environment, and the goal is to develop an interactive model to manage motivation. The chapter contains the concept of developing a motivation model aimed at supporting activity and cooperation of both the students and the teacher. The structure of the motivation model and formal assumptions are presented. The proposed model constitutes the theoretical formalisation of the new situation, when a teacher and the students are obligated to elaborate on a didactic material repository in accordance with the competence requirements. The mathematical method, based on game theory and simulation , is suggested.

5.1 Competence-Based Open Distance Learning

5.1.1 Introduction

During the past 20 years, open distance learning (ODL) has changed considerably, and online teaching and learning have become routine

Open Distance Learning: Fundamentals, Developments, and Modelling
Oleg Zaikin
Copyright © 2023 Jenny Stanford Publishing Pte. Ltd.
ISBN 978-981-4877-55-8 (Hardcover), 978-1-003-13261-5 (eBook)
www.jennystanford.com

practices of many universities (Bates, 2005). The technological development has at the same time given distance education a new appeal for students as well as teachers, and as the amount of distance education has increased, there are a number of definitions connected to the aspects of teaching and learning in an online educational system (Zawacki-Richter and Anderson, 2014). ODL is often used as a synonymou of e-learning and is nowadays considered the most viable means for broadening educational access and is at the same time improving the quality of online educations as well as advocating peer-to-peer collaboration and giving the learners a greater sense of autonomy and responsibility for their own learning (Larreamendy-Joerns and Lainhardt, 2006). ODL is one popular method to provide opportunities as well as to meet the needs of a growing and increasingly diverse student population (Rumble and Latchem, 2003). ODL has a number of potential benefits, for example the ability to overcome the temporal and spatial restrictions of traditional educational settings like space, time, access, etc. (Bates, 2005). Despite all the benefits of ODL, different factors have been identified as crucial to the success of online education. Two of the main factors are the social aspect (interaction among students and teacher) of learning and motivation (Brophy, 2010).

The social aspects of learning are immersed in the social learning context and understanding. The educational experience is constituted through sustained interaction and communication between and among learners, teachers and the learning objects. In ODL this process can be facilitated through asynchronous and synchronous communication and participation (Hrastinski, 2007) and through students' teamwork (Kofoed and Stachowicz Marian, 2012).

Motivation is a key factor in learning and achievement in face-to-face educational contexts (Brophy, 2010) as well as in online learning environments (Hartnett *et al.*, 2011). Traditionally online learners have been understood as independent, self-directed and intrinsically motivated (Hrastinski, 2007), but this is not the case anymore. The motivation of the learner is complex, multi-faceted and sensitive to situational conditions (Hartnett *et al.*, 2011).

Implementation of ODL assumes modification of the traditional organisation of learning processes, and especially the change in designing the structure and content of the specific educational elements, but also the opportunities of relations between students and the teacher seem to be crucial. ODL, combined with the new possibilities and strategies of distance learning, provides a basis for active collaboration between teachers and students, between students and students, as well as between students and the university (Różewski and Zaikin, 2015).

These new possibilities urge universities to consider the new status of the potential students. This means that on the one hand, students must be able to achieve knowledge, skills and competencies connected to the goals in their concrete study programme, and on the other hand, the universities have to provide the student with an active role in the learning process, to motivate him/her to work on an independent study programme under the guidance of a teacher (Barker, 2004).

In this chapter we argue that motivation and social aspects as interaction among teachers and students are essential for a successful learning process in an ODL environment, and our aim is to develop an interactive model to manage motivation. This part contains the concept of developing a motivation model aimed at supporting activity and cooperation of both the students and the teachers in the process of implementing and using an open and distance learning system. The structure of the motivation model and formal assumptions are presented. The proposed model constitutes the theoretical formalisation of the new situation, when a teacher and the students are obligated to elaborate on a didactic material repository in accordance with the competence requirements. It covers two motivation areas, the teacher's and the student's, which describe their interests in knowledge repository development. The interpretation of the cooperation between the teacher and the students is described on the basis of game theory. A game situation in which the teacher plays the role of the game leader and the students can play either in an organised sub-group or can play individually is presented. Analysing the constraints of this cooperation (time, workload), a simulation approach is proposed. The possibility of assessing the various variants of the teacher's work in the different educational situations by using the simulation model and the simulation software is described. Finally we can conclude that there is a possibility to compare the expected costs that the teacher will bear, considering the assumed repository development (working time) with the results achieved in the didactic process (number of students with a high level of competence participating in the repository, number of students with an average level of competence, etc.).

5.2 The Need for Active Cooperation of Students and Teachers in ODL

In terms of a significant reduction or an absence of direct contact between the teacher' s and students' work scope, the way they perform in the teaching and learning processes as well as in research and educational activities varies considerably, and changes occur due to various reasons (Różewski and Zaikin, 2015).

1. In a traditional learning situation, didactic (teaching) material is a support for accumulating knowledge that is seen as necessary for a professional career and that will remain relevant for a long time. This cannot be said for newer educations dealing with new technologies such as computer science, management and business, productions areas, banking and new media technologies (Tencompetence, 2009).

2. In the traditional teaching approach, learning might look like 'the process of intelligent production', except when a student finds out that the learnt knowledge is lacking to be reflected in context and he/she is almost without the skills to use it. The teacher might be aware of the problem and can take it into account during direct contact with the students.

3. The main conventional tool for establishing motivation used by the university systems is knowledge control (tests, exams, grades, etc.).

4. The design and relevance of didactic materials have a secondary role in a traditional teacher's professional role (research is priority one), but it is still the main engine in a student's cognitive learning processes.

Cognitive processes are characterised by high entropy. However, during direct contact with a student the traditional teacher using his/her professional competence can reduce the entropy as a result of certain didactic or pedagogical methods or, in other words, be able to manage the student's cognitive processes within certain limits. In this case the efficiency of teaching/learning management depends on the intensity of the contacts, abilities of students and the teacher's pedagogical and communicative skills. (Skinner, 1953; Różewski and Zaikin, 2015) find that the importance of the role of didactic materials in ODL is greatly increased, as in ODL, teachers have to substitute direct contact and exchange of information between teacher and student. This imposes new requirements to didactic materials not only as information about learning topics but also to see teaching and learning strategies as well as to see the whole learning process as a combination of communication, interaction and knowledge production(Różewski *et al.*, 2011).

The big challenge for ODL teachers is to design and make teaching materials to support ODL. We know that it takes a lot of resources: professional knowledge related to the subject as well as pedagogical skills and, furthermore, knowledge and experience about teaching within the frame of ODL as well as the time for developing and designing new material that is fit for ODL teaching. A solution could be to actively involve the students in establishing knowledge as a common process of teaching/learning activities.

Ontology is widely used for knowledge representation, and there are ontology description languages and programmes to operate them (Gruber, 2009). A more detailed design of didactic materials based on ontology models of subject knowledge is described in ("e-Quality", 2004). Design, preparing and access to the didactic materials in an ODL computer environment (creating, for example, a repository) require specialised knowledge and time from the teacher involved. Therefore the following problems appear when developing ODL: how can teachers be motivated to carry out additional work to establish and maintain didactic materials in the repository up to date? A repository we understand as a virtual or digital library of didactic materials, assigned for learning of a subject or domain area (Kushtina *et al.*, 2007).

One way to minimise the problem is to attract students to supplement the repository with new materials on the proposed teacher topic within a pre-defined ontological domain model. In the ODL situation, the students independently could develop the ontology of the proposed theme and then make it compatible with the conceptual structure and level of details in the ontology of the subject developed by a teacher. Performing this task requires extensive use of highly demanding operations such as generalisation, classification, induction and deduction, but promotes deep mastery of the conceptual apparatus of the subject area. Coombs et al. (Coombs *et al.*, 1970) discuss the advantages of using an ontological approach to learning and teaching. The main advantage of this method is recognised as the development of analytical skills of the students and a systematic vision of wide fields of knowledge objects and their applications. When performing this task the students involved should get first priority when using the teacher's consulting. For students who work together with highly skilled teachers, this activity promotes self-esteem and motivates the self-education activity.

Another way to involve students in active learning is to create a game situation (Shubik, 1991). A distinctive feature of the ontology developed for educational purposes is that the ontology graph contains fragments corresponding to different types of subject knowledge: the theoretical (what is this?) and the procedural (how to do this?). Theoretical vertices describe the semantics of concepts and their relationships, and the procedural ones test tasks associated with the corresponding path in the graph. The project task, to develop the domain ontology with both types of vertices, can be represented as a game with total win and distribution points, depending on student participation and performance.

The overall gain considered is the number of vertices of both types, added to the domain ontology graph, which are stored in the repository. The teacher plays the role of the head of the game, and the students are combined into sub-groups or they can play individually. The teacher's motivation is the possibility of an extension and updating of the repository by independent work of the students. The students' motivation is the joint study of the subject under supervision of the teacher – live chat, cognition through competition, stress reduction compared to traditional testing, choice possibility, etc. The game can be carried out remotely (Kusztina *et al.*, 2010). The final assessment depends on the task complexity, participation in the project and the number of European Credit Transfer System (ECTS) points. The students will progress from a simple task to a more complicated task due to the mechanism of motivation, created by a teacher.

5.3 Interpretation of the Motivation Model of Learning Processes

In traditional teaching, using traditional didactic material design and remote technologies, the average activity of the students' involvement in the learning process is weak, and the accumulation of material in a repository for subsequent use is small, but the teachers' time costs are lower than in the following two examples:

1. Independent development and updating of the repository by students and use of it as a didactic material for developing knowledge in ODL can raise the activity of the students, the costs of the teachers' time will increase significantly, but it is justified by the possibility of obtaining results for further improvement and modernisation of the learning processes.

2. Students' teaching – how to use the ontological approach for representation and development of knowledge in any domain area and organisation of controversial games as a way of self-learning in a given subject area – might result in a significant rise in students' activity. The time costs of the teacher are comparable to the first case. But in the second case the students not only acquire knowledge but also competence, and the teacher in return gets new improved didactic material to be used in the ontological approach and in the games tools.

For evaluation and assessment the teacher can use the linguistic scales to evaluate the proposed alternatives. In each situation the teacher can give preference to any of these alternatives, as each of them is connected with its various temporal and intellectual costs as well as other preferences. The teacher's intellectual cost is related to checking and evaluating complex tasks. The more students choose complex tasks, the more intellectual cost the teachers have to spend for their checking and evaluation.

One of the impact factors of the teachers' choice is the evaluation to get the final result at the end of the learning in the form of specific indicators such as:

- The average index of student performance.
- An index of students' activity and independence of choosing types of tasks, for example the traditional way of learning and testing, the work together with the teacher on the use of the ontological approach for repository development and an engaging in joint or individual learning projects
- The index of repository filling by didactic materials is created on the principle of 'actual domain ontology' plus the base of learning projects as a kind of 'best practice'.

The availability of such data allows for further research, not only in teaching methods, but also in the field of artificial intelligence and the development trends of corresponding knowledge domains. This, in turn, is a problem raised for discussion in a lot of conferences, publications and journals (Novikov, 2012).

The modern system of assessment of the teacher's professional activities requires wide participation in international scientific research. In the proposed game this can play an important role for evidence-based evaluation of the learning process.

The scoring system used for evaluation of the students' work is the teacher's prerogative and can be addressed to the student groups in the beginning of the learning process, but the student's choice will also depend on his/her personal preferences:

- Leadership skills
- The possibility of increasing the total average diploma score
- The possibility of participation in individual grants due to personal contribution to the 'best practices'

The proposed approaches to learning allow one to realise motives and preferences of both students and teachers, so we can form, respectively, two motivation models and examine their mutual influence.

A motive (the reason of action) is a consciously understood need for a certain object, position, situation, etc. Therefore we can state that the motive comes from a requirement, becomes its current state and leads to certain actions (Anderson, 2000). During the realisation of the motive as a need, the student is dealing with a decision situation, which is a motivational factor (Tuckman, 2007). A motive is formed as a conscious request of something (object, position, situation, etc.). Further, it is transformed to an actual need and becomes the main factor in the subsequent action. While the chain 'inquiry - motive – action' has been formed and implemented, we can deal with the problem of difficult choices at each stage. As several motives can induce a particular act, several different objects can transform to one motive.

Choice-making is a cognitive process, and it cannot be directly observed. This means that the relationship between a person's preferences in motives when choosing the result of his/her actions is difficult to define. Nevertheless, the choice of motive can be evaluated by registering the end outcomes.

A motivation model in the described learning situation can be represented as a game scenario, in which the activities of the teacher and students will respond to their preferences in choosing the above alternatives. The proposed model has the following assumptions:

- The elements of the described chain and its alternatives are pre-defined.
- The choice of a job in the game is made once.
- For each task, which the student selects, the degree of its complexity and the corresponding number of ECTS points is known.
- The content of the selected task can be displayed on an ontology graph of the studied subject as a set of vertices with the theoretical and procedural knowledge, which constitute a certain portion of competence.
- Students and the teacher will have the possibility to know the results of the partners' choices in the game.

The research discussion about the motivation model can be addressed to different areas of the educational system. The conducted analysis of information processing in judgmental tasks allows to prepare a cognitive-motivational model of decision satisfaction (Keller, 1999). In the proposed model confidence serves a role of a major contributing factor of learning motivation. However, more detailed investigation proves that the motivation is a set of several components. The attention, relevance, confidence, and satisfaction (ARCS) motivation model (Miller and Brickman, 2004) is based on four-factor theory. The student's motivation is hooked up with the student's attention, relevance, confidence and satisfaction. The ARCS model also contains strategies that can help an instructor stimulate or maintain each motivational element. Other researchers shown that personally valued future goals are the core for motivation (Phalet *et al.*, 2004). Moreover, the cultural discontinuities and limited opportunities in the students' learning environment may weaken the motivational force of the future (de Sevin and Thalmann, 2005).

The form and content of the motivation model is also strongly dependent on the object to be motivated and the environment where motivational action takes place. On the one hand, the motivation model can be designed for an artificial or a human object. (DeVoe and Iyengar, 2004) propose a motivation model for virtual humans such as non-player characters. The motivation model based on overlapping hierarchical classifier systems, working to generate coherent behavioral plans. On the other hand a different environment creates individual needs for the motivation model. Such a situation is caused mainly by multicultural differences (Kushtina *et al.*, 2007).

5.4 Statement of the Motivation Problem in a Particular ODL Situation

In the proposed game scenario each participant has the possibility of free choice among the available alternatives. The main result of the game is to develop an ontology graph of the studied subject by means of adding new vertices connected with the existing ones. These new vertices submit new theoretical knowledge and/or skills acquired in task-performing. The goal of the teacher is to attract as many students to this kind of tasks under the following constraints: the total number of hours assigned for the subject in scheduling and the time interval allotted for the game.

Each student personally defines the result of the game in the range of the following features: from the lowest allowable assessment on the subject (in the traditional way of studying) to the highest result for active and successful participation in the project and 'best practices'. A teacher distributes the supervision time among students at his/her discretion, but at the condition of a minimum mandatory quota for each student. In this situation it may be assumed that a teacher is interested to spend the most of his/her time and intellectual resources on joint work with students. They will contribute to the updating and expansion of the repository, which provides the basis for compiling the results in the studied subject, with an option to use grants or publications. This aspect motivates the teacher to advance students to more complex work and to spend more resources on their performance. In this case there is a possibility to use the results in the teacher's/students' common research. Thus, the proposed scenario allows the teacher and each student to formulate the conditions to choose their strategies in the learning process on the basis of individual preferences.

The teacher's objective will be to maximise the possible extension of the repository through the involvement of a larger number of students for competitive performance of complex project tasks under the restrictions on the total time of supervision.

Considering students' preferences , they may be defined within two extreme cases:

a) The students prefer to perform the simplest tasks, minimising their time and intellectual costs and getting the lowest possible scores. In fact, these students don't participate in the repository extension.

b) The students prefer to participate in the repository extension of knowledge acquired in the project task under restrictions to their own time and intellectual resources. Every student intends to use the maximum teacher's time of supervision, competing for it with other students in this group.

In the first case, the teacher and students with minimal effort will complete the learning process, but indicators, such as average achievement and filling of the repository will be low, although satisfying the low boundary conditions (formal compliance schedule and personal quota for supervision).

In the second case, the teacher and students will spend far more effort, but a great part of successfully completed tasks will be included into the repository and the average achievement will be higher.

At the same time, in the second case, learning will be aimed at understanding not only the theoretical material but also the ability to operate it (i.e. mastering of didactic material at the level of competence).

Thus, the aim of the game is the maximum extension of repository with original results of the teacher's and students' joint work under restrictions on their time resources and in compliance with preferences of all participants of the game.

The motivation problem is considered in particular educational situations when we are dealing with special didactic material for particular participants of the learning process. For this we will use the term 'learning situation', which is represented in Fig. 5.1.

The formal model that is describing the competence-oriented learning process (Kushtina *et al.*, 2009) has the following structure.

5.4.1 Basic Components of the Learning Situation

a) Participants of the learning process $\{N, S\}$, where
    N is the teacher,
    $S = \{s_j\}$ is the set of students/project team, where
    $j = 1, 2, ...j^*$ is the index of the student.
The process of students' arrival to an ODL system can be described by a pattern

$$\Pi(S) = \{\chi, \lambda\},$$

where χ is the distribution law, λ is the intensity of arrival,

b) Ontology graph of the domain area G^D

$$G^D = \{W^D, L^D\},$$

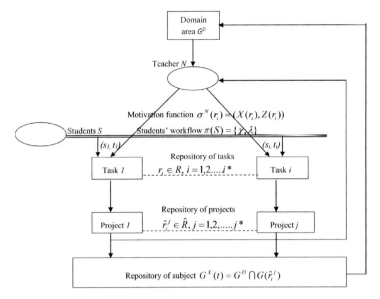

Figure 5.1 Process of choice and execution of tasks/projects by students (learning situation).

where
> W_D are nodes/concepts/learning objects of the graph,
> L_D are graph edges (relations between concepts),

c) Ontology graph of the course G^C

$$G^C = \{W^C, L^C\},$$

where $W^C \subseteq W^D, L^C \subseteq L^D$

d) set/repository of the tasks R

$$R = \{r_i^k\},$$

where
> r_i^k is the task 'i' consisting for competence portion 'k',
> $i = 1, ..., i^*$ is the index of the task,
> $k = 1, ..., k^*$ is the index of the acquired competence/competence portion

e) Parameters of tasks $\Pi(r_i^k)$

$$\Pi(r_i^k) = \{Q(r_i^k), A(r_i^k)\},$$

where

$Q(r_i^k)$ is the level of complexity of a task, which can be expressed in a binary scale

$$Q(r_i^k) = \begin{cases} 1, & \text{if task } r_i^k \text{ is complex one} \\ 0, & \text{otherwise} \end{cases}$$

or in numerical scale (e.g. number of concepts from set W^D included in the task),

$A(r_i^k)$ is the topicality of the task, characteristic of a task, assigned by the teacher, which can be expressed with a binary scale,

$$A(r_i^k) = \begin{cases} 1, & \text{if task } r_i^k \text{ is chosen by the teacher} \\ & \text{to develop the repository} \\ 0, & \text{otherwise} \end{cases}$$

The teacher places an up-to-date, properly solved task in the task repository R.

f) Repository of the projects assigned by teacher for saving of subject knowledge X

$$X = \{\hat{r}_i^k\},$$

where $\hat{r}_i^k$ is the executed task included in the repository

g) State of the repository X in time t, $G_X(t)$

$$G_X(t) = \{W_X(t), L_X(t)\},$$

where

$W_X(t) \subseteq W^C$ is the set of vertices of graph $G_X(t)$,
$L_X(t) \subseteq L_C$ is the set of arcs/edges of graph $G_X(t)$,
$t \in [t_0, T_c]$ is the time interval assigned for acquiring of the competence

5.4.2 Decision Parameters

a) The student's decision parameter: binary function of the task choice $y(r_i^k, s_j)$

$$y(r_j^k, s_j) = \begin{cases} 1, & \text{if student } s_j \text{ chooses task } r_i^k \\ 0, & \text{otherwise,} \end{cases}$$

b) The teacher's decision parameter: binary function of assignment of the task/project to repository $\delta^N(\hat{r}_j^k)$

$$\delta^N(\hat{r}_i^k) = \begin{cases} 1, & \text{if project/task } p_i^k \text{ is selected for repository} \\ 0, & \text{otherwise,} \end{cases}$$

5.4.3 Criterion and Objective Functions

Let's define the following variables:

a) Ontology graph of the task/project $G(\hat{r}_j^k)$

$$G(\hat{r}_j^k) = \{W(\hat{r}_j^k), L(\hat{r}_j^k)\}, \tag{5.1}$$

where:
$W(\hat{r}_j^k)$ is the set of the vertices of the graph, $W(\hat{r}_j^k) \subseteq W^C$
$L(\hat{r}_j^k)$ is the set of the arcs of the graph, $L(\hat{r}_j^k) \subseteq L^C$

b) Knowledge gain under replacement of task/project $\hat{r}_i^k$ in repository $X, \Delta_X(\hat{r}_i^k)$

$$\Delta_X(\hat{r}_i^k) = G_X(t) \cap G(\hat{r}_i^k) = \{W_X(t) \cap W(\hat{r}_i^k), L_X(t) \cap L(\hat{r}_i^k)\} \tag{5.2}$$

c) Numerical characteristic of knowledge gain in repository $|\Delta G_X(\hat{r}_i^k)|$

$$|\Delta G_X(\hat{r}_i^k)| = \delta^N(\hat{r}_i^k)|W^C \cap W(\hat{r}_i^k)|, \tag{5.3}$$

where $|W^C \cap W(\hat{r}_i^k)|$ is the number of common vertices in the ontology graphs G^C and $G(\hat{r}_i^k)$

d) Teacher resources expenditures $Z^N(\hat{r}_i^k)$, for example consultation time for task/project $\hat{r}_i^k$ executing

$$Z^N(\hat{r}_i^k) = \alpha_N|G(\hat{r}_i^k)|, \tag{5.4}$$

where α_N is the weight coefficient of teacher expenditures.

5.4.3.1 Objective function of the teacher

a) Total gain of repository X on the interval of acquiring of the competence $\Delta G_X^\Sigma(0, T_0)$

$$\Delta G_X^\Sigma(0, T_0) = \sum_{i=1}^{i^*} \sum_{k=1}^{k^*} \sum_{j=1}^{j^*} \delta^N(\hat{r}_i^k) y(r_i^k, s_j) \left(G_X(t) \cap G(\hat{r}_i^k)\right) \tag{5.5}$$

b) Numeric characteristic of total knowledge gain in repository $|\Delta G_X^\Sigma(0, T_0)|$

$$\Delta G_X^\Sigma(\hat{r}_i^k) = \sum_{i=1}^{i^*} \sum_{k=1}^{k^*} \sum_{j=1}^{j^*} \delta^N(\hat{r}_i^k) y(r_i^k, s_j)|W^C \cap W(\hat{r}_i^k)|, \tag{5.6}$$

where $|W^C \cap W(\hat{r}_i^k)|$ is the number of common vertices in ontology graphs G^C and $G(p_i^k)$.

c) Total expenditures of the teacher on interval of acquiring of the competence Z_N^Σ

$$Z_N^\Sigma = \sum_{i=1}^{i^*} \sum_{k=1}^{k^*} \sum_{j=1}^{j^*} \alpha_N y(r_i^k, s_j)|W(\hat{r}_i^k)|, \tag{5.7}$$

where

α_N is the weight coefficient of the teacher's expenditures,
$|W(p_i^k)|$ is the the number of vertices in ontology graph of task/project $G(\hat{r}_i^k)$.

d) Objective function of the teacher Φ^N: total gain of repository X on the interval $[t_0, T_c]$ of acquiring of the competence, taking into account expenditures of the teacher

$$\Phi^N = |\Delta G_{\tilde{X}}^{\Sigma}(0, T_0)| - Z_N^{\Sigma} =$$

$$= \sum_{i=1}^{i^*} \sum_{k=1}^{k^*} \sum_{j=1}^{j^*} y(r_i^k, s_j) \{\delta^N(\hat{r}_i^k)|W^C \cap W(\hat{r}_i^k)| - \alpha_N|W(\hat{r}_i^k)|\} \qquad (5.8)$$

5.4.3.2 Objective function of the student

Objective function of student $\Phi(s_j)$: number of ECTS points, taking into account expenditures of the student for execution of task/project $\hat{r}_i^k$

$$\Phi(s_j) = \sum_{i=1}^{i^*} \sum_{k=1}^{k^*} y(r_i^k, s_j) \gamma^N(\hat{r}_i^k)|W^C \cap W(\hat{r}_i^k)| - \beta^S|W(\hat{r}_i^k)| \qquad (5.9)$$

where
$\gamma^N(\hat{r}_i^k)$ is the number of ECTS points assigned by the teacher for the task/project $(\hat{r}_i^k)$,
$|W^C \cap W(\hat{r}_i^k)|$ is the numerical characteristic of gain of repository for task/project $(\hat{r}_i^k)$,
$|W(\hat{r}_i^k)|$ is the numerical characteristic of complexity of task/project $(\hat{r}_i^k)$,
β^S is the weight coefficient of student's expenditures.

5.4.3.3 Constraints

a) Summary resources (time-related, technical, didactic, staff) offered to students for solving tasks:

$$\bar{Z}_0 = \sum_{s_j \in S} \bar{z}(r_i^k(s_j)) y(f_i^k(s_j)) \le Z_{\Sigma}^N, \qquad (5.10)$$

where
$\bar{z}(r_i^k(s_j))$ are the resources appointed to student S_j for solving task r_i^k,
$y(r_i^k(s_j)) = \{1, 0\}$ is the binary function of choice the task r_i^k by student S_j,
$\bar{Z}$ are the summary resources for the subject lead by the teacher.

b) Calendar interval $\tau \in [0, T_0]$, appointed to students for choosing and solving tasks

$$\min_j \underline{\tau}(r_i^k(s_j)) \geq 0, \quad \max_j \overline{\tau}(r_i^k(s_j)) \leq T_0 \qquad (5.11)$$

where $\underline{\tau}(r_i^k(s_j))$, $\overline{\tau}(r_i^k(s_j))$ are appropriate moments to start and end solving task r_i^k, respectively, by student S_j.

5.5 Motivation Model Interpretation in Terms of Game Theory

The interpretation and solution of the developed model can be conducted on the basis of game theory, which allows studying the activity of a system, depending on the players behavior (Gubko and Novikov, 2002).

The proposed model refers to the class of games with a defined number of steps and full information about participants' activities in real time. The game has an arbitrary sum of participants' wins:

- The win of the teacher is the gain of knowledge in the repository.
- The win of the student depends on his/her strategy: maximal number of points for a task solved.

The equilibrium is obtained as a result of a *dominant strategy*, which compared to other strategies gives the game participants the possibility to obtain their maximal win regardless of the actions of other participants.

Using game theory terminology, the motivation model can be seen as an incentive task, where motivation management signifies direct rewarding of an agent (student) for his/her actions. The formulated model is consistent to a *multi-agent two-layer incentive system* that consists of one centre (teacher) and *n* agents (students). The strategy of each agent is to choose an activity, and the centre's strategy is to choose a motivation function, that is, a relationship between the win of each agent and his/her actions.

Let us denote *M* as a set of acceptable motivation methods and $Y(\sigma)$ as a set of game solutions (strategy of agents having balance in their motivation method σ). Management (motivation) effectiveness means obtaining the maximum value of the objective function $U(\sigma)$ on an appropriate set of game solutions.

$$U(\sigma) = \max_{y \in Y(\sigma)} f(\sigma, y), \qquad (5.12)$$

where σ is a motivation function of the centre and y is a binary argument of agent's choice.

The task of optimisation is about searching for an acceptable motivation function with maximum effectiveness:

$$\sigma \in Arg \max_{\sigma \in M} U(\sigma). \qquad (5.13)$$

When solving the model, algorithms described in (Novikov, 2012; Gubko and Novikov, 2002) can be used. At the moment of the decision making (motivation function for the centre and student's decision function for the agent), the objective functions and acceptable actions of all participants are known.

The sequence of steps in the game is as follows:

1. The centre has the right of the first move, when it chooses a motivation function, before the agents choose activities that optimise their objective functions. The centre's (teacher's) objective function Φ^N is the difference between expected income X (gain of repository) and the summary reward paid to the agents (sharing one's own resources Z_N^{Σ}).

$$\Phi^N = X - Z_N^{\sigma} \qquad (5.14)$$

The centre's choice of a motivation function takes place under the condition of uncertainty and foreseeing random characteristics of the basic students' knowledge and parameters of the process of their arrival. Therefore the centre can use the principle of minimal guaranteed result and compensatory function of restitution, which means a lack of the centre's income. In that case the centre's objective function has a negative value and represents the minimal costs of the centre when working with agents.

$$X = 0, \quad \Phi^N = -Z^N = min \tag{5.15}$$

Therefore the centre creates a motivation function under the condition of a minimal guaranteed result, $-Z^N = min$, which cannot exceed the admissible centre's costs $-Z_\Sigma^N$.

2. Agents choose their strategies independently and do not exchange information or wins, which signifies that we are dealing with a relational dominant strategy (RDS). By the objective function of each agent Φ^S, we understand the difference between his/her reward obtained from the centre γ^N and the losses connected to solving the task w^S.

$$\Phi^S = \gamma^N - w^S \tag{5.16}$$

If the income of the centre $X = 0$ and the reward of each agent only compensate his/her estimated costs $\gamma^N = w^S$, then the objective function of each agent

$$\Phi^S = 0.$$

3. On the basis of the result of the agents' selection, the centre can adjust the function of motivation to achieve its maximum value of the objective function. In addition the centre can verify restrictions on the total resources provided by agents.

$$\Phi^N - X - Z_N^\Sigma = max$$
$$Z_N^\Sigma \leq \hat{Z}_N \tag{5.17}$$

4. The agent makes the final selection of a task on the basis of his/her preferences, while maximising its objective function.

$$\Phi^S = \gamma^N - w^S = max \tag{5.18}$$

Figure 5.2 shows the functioning schema of the described scenario of filling the task repository in an ODL system.

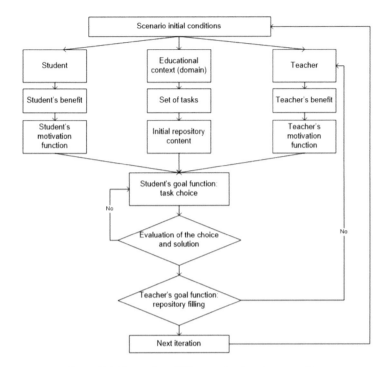

Figure 5.2 Scenario of filling and using the repository.

The learning situation can be characterised by a proper ontology, a part of which is conformant with the task ontology (in the sense of the amount and depth of the concepts being used).

The teacher's interest is in maximising the level of repository filling with tasks of different complexity for every considered educational situation in any given domain. The criterion regarding placing a task in the repository is decided by the teacher on the basis of the complexity level, graphical quality, language correctness, etc. The possibility to realise the teacher's interests is limited by his/her resources, considering time in the quantitative and calendar aspects and other informal preferences.

The teacher's motivation function concerns maximising the coverage of the learning situation with tasks (prepared and solved) with defined topicality of the task's subject from the teacher's point of view and the individual resources he/she is prepared to appoint to a student for solving a certain task (consultation time, access to scientific material, equipment, etc.). The teacher defines his/her motivation function before students gain access to choosing tasks to solve this function has to be known to them.

The student's interests depend on individual preferences and can be divided into two groups. The first group of students is interested in achieving the minimal acceptable success level, meaning meeting only the basic requirements for obtaining a positive grade for solving the task (low complexity of the task, minimal acceptable grade), while saving the maximal amount of their time. Such task-solutions will not be placed in the repository. The second group of students is interested in filling the repository with the maximal possible success, meaning solving tasks of high complexity that will later be placed in the repository as their own copyrighted part of didactic material.

The student's motivation function considers obtaining the maximum level of fulfilling one's interests during choosing and solving the task, with given constraints regarding time (one's own and the teacher's) and the way of grading the resulting solutions.

The statements made above show that both motivation functions depend on the complexity level of a task and have common constraints on time resources. Supplementing the repository with new tasks can be interpreted as accrual of the knowledge resource. Increasing motivation of both students and teachers increases and quickens the accrual of this resource.

Therefore, the motivation model in an ODL system (regarded as simply 'motivation model' from now on) should include parameters describing activities of each interested side: the student and the teacher. The measure of successfulness of their cooperation is the accrual of knowledge in the repository, which can be evaluated through the intensity of its filling with properly solved tasks. When developing a motivation model, one has to consider a very important factor: the stochastic character of the students' arrival, which is mainly a result of the individual education mode and stochastic character of students' motivation parameters. The motivation model regulates the process of a student choosing a task to solve within the scope of a certain subject on the basis of his/her own motivation function and with consideration of teacher's requirements and preferences.

The entire process, from the moment of formulating tasks to the moment of evaluating them and placing in the repository, followed by creating a new set of tasks waiting for the next group of students prepared to solve them can be described by a game scenario. The presented scenario is universal in every learning situation aimed at obtaining not only a portion of knowledge but also a competence based on it. Modelling a game scenario requires formulating motivation and goal functions of the game participants with regard to the repository filling.

5.6 The Procedure for the Acquisition of Personal Competence

5.6.1 The Algorithm Acquisition of Competencies

Within psychology there is a distinction between the mastery of operation and the mastery of theoretical knowledge and acquisition of competencies. Based on (Woolfolk, 2012), we will try to present the process of acquiring competencies based on a certain level of domain knowledge. The aim of the execution procedure is:

(1) Acquisition of competence training for students: the level of theoretical knowledge, the nature of the competencies during a training session

(2) Collection of statistics, allowing for clarification of the individual process management model of teaching in the ODL system

The training needed to get the student through the mechanisms of the repository's virtual lab with the set of triplets: *part of the description fields – a typical task – a typical solution*, and the corresponding test task that is used to describe knowledge as an extended ontological model.

The task of testing should be interpreted in the terms contained in the proposed triplet. For proper identification of test task, a disclosure of knowledge necessary to prepare a solution will be needed. The student's solution can be compared with the solution typically stored in the triplet. The measurement of teaching is not just about collecting statistics of personal data, but is also about analysing the interpretation of test tasks.

The implementation of a training session can be calculated using the following indicators:

a. The percentage of correctly interpreted jobs
b. The percentage of correctly identified tasks
c. The percentage correctly associated pairs of a typical task–typical algorithm
d. The number of modules placed in the repository of knowledge
e. The number of sessions needed to perform a specific task
f. The number and complexity of the problems included in one job

The described indicators allow one to measure teaching, depending on skills of mastering theoretical knowledge and skills of its use in solving design tasks. The procedure includes several steps, and the algorithm of its implementation depends on the software environment in which the repository will be created (Fig. 5.3).

Input data of the procedure:

1. Scope: subject/topic of teaching.
2. The model of theoretical knowledge areas: a hierarchically ordered the basic concepts of the semantic network (graph).
3. The reference model to enable the use/assumption of the taxonomy studied during training problems; the presentation depends on the specific areas (S_p).
4. Repository solved tasks: typical task – a typical algorithm $\sum(Z_i, A_i)$, 1, ..., i is the number of tasks.

At the start of the session the repository can be empty or filled, depending on the topic and the student's personal learning goals.

The procedure for the acquisition of competencies during structured training is done according to the following steps:

1) *Analysis of the research problem.*

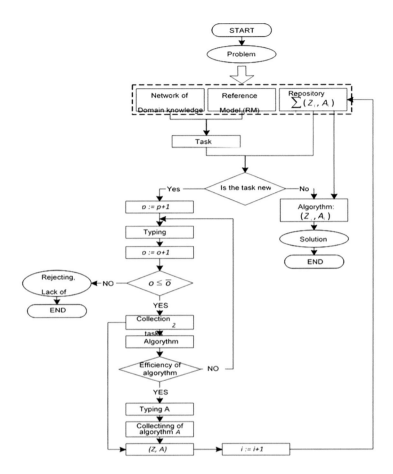

Figure 5.3 Algorithm acquisition of competencies during structured training.

Determining whether the issue belongs to specific areas that will allow one to interpret the problem and to present it in terms of a particular model, taking into account existing knowledge of taxonomy – the formulation of the input problem.

Depending on the educational situation the first step can be done in two ways:

a) Independent formulation of the task by the student on the basis of the ontological model G_p and reference model of typical tasks

      b) Analysis and identification of the tasks set by the supervisor of the project

2) *Analysis and systematising of experience.*
A comparison between the input tasks with the tasks placed in the repository. The result is to determine the path that must be used to solve the problem either the way to develop an algorithm.

3) *Typing of input tasks.*
Preparation of the task passport in the repository language (meta-information in an XML document).

4) *The collection of the input task passport* in the working memory of the training session.

5) *Developing of own algorithm for solving the task.*
The algorithm can be described using standard pseudocode or presented as a simulation task.

6) *Execution of the algorithm.*
The input data should be selected directly from the text of the analysed problem or pulled during its interpretation.

7) *Evaluation of the effectiveness of the algorithm.*
At this stage the interpretation of results is made in the context of the output of the algorithm (in terms of the solved problem).

8) *Typing of the developed algorithm.*
Preparation of the solution algorithm passport in the repository (meta-information in an XML document).

9) *The collection of the algorithm passport* in the working memory of the training session.

10) *Preparing the knowledge module in the form of a record in the repository.*
At this stage the teacher has to fill in a repository form, including track keywords of the domain knowledge, matching the given problem, the task passport and the passport algorithm.

11) *Supplementing of the existing repository.*
The required level of the supplement depends on the subject, purpose and stage of teaching and must be given by the teacher to each student.

The described training is one of the traditional tools to collect statistics in the course of self-education students. However, the proposed structure for the fulfillment of the repository has several advantages:

a) Corresponds with the cognitive meaning of the competency structure
b) Represents a system approach to a knowledge ontological model
c) Enhances the ability to structure theoretical knowledge and connect it with the results of the student's own experience

A similar idea is used in simulators, the aim of which is to teach the use of complex equipment repair and machinery. The proposed method of the domain area describes the structure and content of the repository and allows students to master the complex theoretical knowledge while the teacher is absent. In this case, a chain 'typical task - typical solution' is represented at the level of a mathematical reference model, which depends on the specific subject and the purpose of education.

This procedure is a model skeleton in the course of training related to a specific topic. The ontological model, repository and task should be completed with relevant teaching materials. An additional variant of the described approach to the structure and contents of the repository is associated with the formation of a *virtual laboratory*. This is designed to conduct simulation experiments, when the assumption of the task simulation, execution and interpretation of the results of the experiment needs the cooperation of experts located at a distance.

5.6.2 Ontological Graph of the Course Consistent with the Structure of Competence

The first ontologies were built for the needs of engineering knowledge, and they began to emerge in the 1980s. Intensive works on ontology resulted in the development of its various categories. Proposed by Guarino (Guarino, 1998), the division includes:

- General ontology (a top-level ontology)
- Domain ontology
- Task ontology
- Application ontology

Each of these ontologies has a different level of generalisation.

Domain ontology describes the domain area knowledge characteristic, for example medicine, pharmacy, law and music. The terms used in domain ontology are the result of specialisation of concepts defined in general ontology.

Task ontology describes the dictionary of a particular task or activity, for example diagnosing, scheduling and specialising terms stemming from general ontology. In contrast to domain ontology, in this case, for solving the problem concepts from various fields can be used.

Application ontology includes the concepts that are required for the description of knowledge for individual applications. It specialises the concepts of domain and task ontologies for a given application.

For example, an ontological graph of the subject/course 'Discrete mathematics', provided for the specialty 'Computer Science', is presented in Fig. 5.4.

The *practical part of the course* includes a number of tasks. There are two kinds of tasks: simple and complex. They differ in the labour complexity of implementation and the number of ECTS points, which the student will get for correct execution of the task. In general the teacher can define any number of classes of tasks, which differ in the degree of complexity.

From the point of view of learning objectives, the whole group of students can be divided generally into two extreme groups. For the first group of students (interested in achieving the minimal acceptable success level) the motivation function is a just monotonously falling function of a discrete argument, meaning the task complexity. For the second group of students (interested in filling the repository with the maximal possible success level) the motivation function is a monotonously rising function of the same argument.

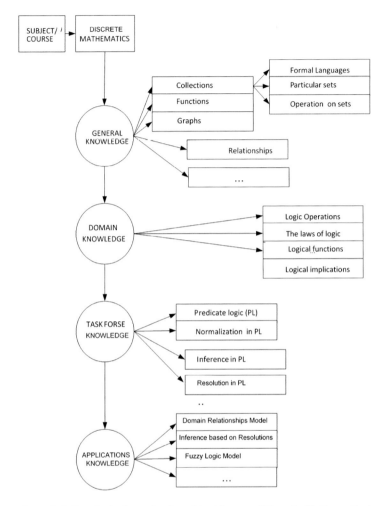

Figure 5.4 Ontological graph of a subject/course 'Discrete Mathematics'.

5.6.2.1 Goal function of the student's task choice

Under effectiveness of the decision made we understand maximal satisfaction of the student's and teacher's interests with the maximal summary motivation function. The form of satisfying the teacher's interests is placing in the repository a properly solved task of a significantly high level of complexity. The form of satisfying the student's interests is minimal summary time costs, while obtaining a high grade, which also depends on the complexity level of the task. The period of filling the repository is limited by a calendar interval, depending on the educational situation.

The fact of making the decision can be described by a binary argument,

$$y_i^j = \begin{cases} 1, & \text{if student } s_j \text{ chooses task } r_i \\ 0, & \text{otherwise,} \end{cases}$$

Then, the goal function has the following structure:

$$\Phi(y_i^j) = \alpha \sigma^T + \sigma_j^S = \max_Y, \tag{5.19}$$

where

$Y = \{y_i^j\}, i = 1, 2, ..., i^*, j = 1, 2, ..., j^*,$

σ^T is the teacher's motivation function,

σ_j^S is the motivation function of each arriving student s_j,

α is the waging coefficient.

Both elements of the goal function depend on the same argument: task complexity level $Q(r_i)$. As shown in Fig. 5.5, the element σ^T is a monotonous rising function of argument $Q(r_i)$, while σ_j^S in dependence on the kind of student is a monotonously falling or rising function of the same argument $Q(r_i)$.

Motivation of students in both cases is described as a linear function. Depending on whether students are ambitious or non-ambitious, this function is respectively increasing or decreasing.

Identifying the teacher's motivation assumes that the function of this motivation will be built for each group of students. Its shape will depend on which group of students the teacher will work with at a given time. The curve describing the *motivation function of the teacher* in the general case is non-linear and convex. In the case of non-ambitious students, the teacher's motivation function is increasing due to the need for continuously motivating students to solve more difficult tasks and requiring more time. Otherwise, the students will be ambitious and strongly determined to achieve a high score, which will be a decreasing function. Due to existing restrictions the teacher must inhibit the motivation of students. The distribution of this function is related to the time constraints that are connected with the teaching process.

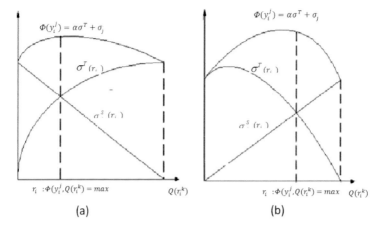

(a) (b)

Figure 5.5 Extreme cases of the task choice function on the basis of motivation and preferences functions: (a) group of non-ambitious students (interested in achieving the minimal acceptable success level) and (b) group of ambitious students (interested in achieving the maximal acceptable success level).

The nature of the motivation function indicates that it is non-linear. The shape of this function results in the rules for summarising the functions of teacher and student motivation. Both components of the objective functions depend on the same argument r_i^k as opposed. The summarising function of student motivation reaches its maximum value when making the optimal choice task r_j^* ($Q(r_i^*)$ – the optimal level of the task complexity).

5.7 The Linguistic Database as a Tool Supporting the Level of Student Motivation

The recognition of teacher and student motivation is the basis to determine whether with the existing restrictions on the competence acquisition process, they can achieve the required competence level and at what cost. Checking these constraints is done using simulation.

The basis to estimate the input value of the simulation model is the expected activity of participants resulting from their motivation. Therefore, recognition of this motivation is necessary to perform the simulation experiments (Zaikin *et al.*, 2014).

Teacher motivation identification is a task so easy that he/she is one and could recognise his/her motives. A more complicated issue is to assess the students' motivation, because there are many students and each of them has a different level of motivation. Therefore, it is proposed to use the mechanism to determine the motivation in the form of a *linguistic knowledge base*. In the presented example it includes a set of student's motivation parameters, which can be based on the experience of the teacher. A linguistic database to support the process of identifying students' motivation is presented in Fig. 5.6.

The structure of a linguistic database to support the process of identifying students' motivation is presented in Fig. 5.7.

The supporting tools that determine the level of student motivation is the linguistic knowledge base (Zaikin *et al.*, 2014). For selected factors influencing the motivation we developed distribution functions of linguistic quantifiers. Two distribution functions of the features are presented in Fig. 5.8.

Individual aggregation allows categorising students within the presented space of the motivation function. If each student determines proper motivation and the student is placed in the

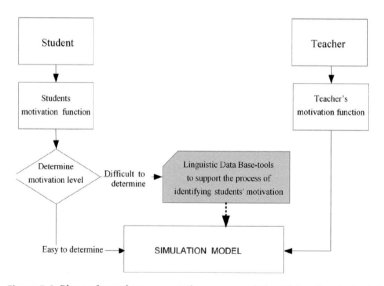

Figure 5.6 Place of a tool to support the process of identifying the students' motivation.

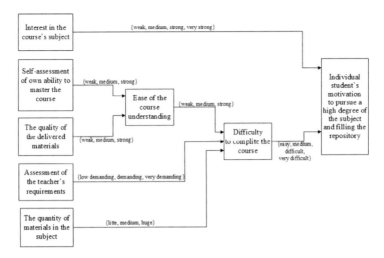

Figure 5.7 Structure of linguistic database for the purpose of identifying the student's motivation.

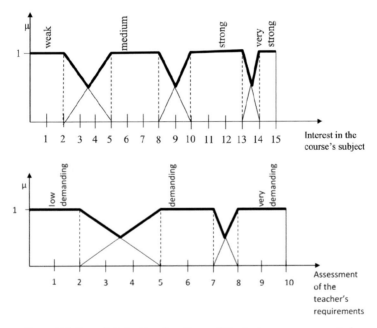

Figure 5.8 Two distribution functions of the features are presented.

depicted space, this information may be used to estimate the initial setup of simulation, for example forecasting of students' service time or forecasting the probability of outputs from the model for different degrees of competence development.

Chapter 6

Collaboration Modelling in Open Distance Learning

6.1 Representation of Competence as a Classical and Fuzzy Set

6.1.1 Introduction

The ability to create a project team and manage its knowledge is commonly recognised as one of the most important qualities of a knowledge and innovative organisation. The management of knowledge and skills, as well as the management of project competencies, has turned out to be essential factors influencing performance of a project process. In this chapter, the author presents the method and tools for competence modelling for project knowledge management. On the basis of this and also using a fuzzy competence model the cost estimation model in the form of a group competencies expansion algorithm is proposed. The model focuses on the cost of project team competence expansion caused by the project development process.

Open Distance Learning: Fundamentals, Developments, and Modelling
Oleg Zaikin
Copyright © 2023 Jenny Stanford Publishing Pte. Ltd.
ISBN 978-981-4877-55-8 (Hardcover), 978-1-003-13261-5 (eBook)
www.jennystanford.com

6.1.2 Representation of Competence as a Classical Set

The basics of mathematical competence models were created in the early nineties by Yu in his work entitled '*An Integrated Theory of Habitual Domains*' (Yu, 1990). Analysing the psychological theories, behavioural and cognitive, it can be said that each person has specific ways of reacting to stimuli in the form of so-called '*collaborative behavioural patterns*' (Simon, 1974). In the course of project development, each participant of the project acquires a set of typical ways of responses, reasoning and decisions as well the knowledge and experience on which they are based. A set of such typical responses, thinking and perception was named *habitual domain* (HD). Then it can be concluded that the HD of the individual determines how it functions in certain situations (Yu, 1990). The concept of the HDs has become the basis for creating *mathematical models of competence* presented as a set, whose elements are the portions of knowledge, skills and abilities owned by a person (Kushtina, 2006).

As mentioned above, the HD is a stable, settled in the mind of the human, pattern of behaviour, response and decision-making (Yu, 1990). According to (Yu and Zhang, 1990) for every project denoted by P a set of competencies can be defined consisting of the knowledge and skills necessary to achieve a satisfactory solution to the project $Tr(P) = \{c_i\}$ (called *truly needed competence set*). A team of persons/students S intending to solve the project P can in their HD have the patterns of procedures for this project, which define the set of competencies $Ck(S)$. We can conclude that for team S to effectively solve the project P, its set of competencies $Ck(S)$ should as far as possible cover a set of competencies required $Tr(P)$. When the cover is unsatisfactory the persons of the team should extend their set of competencies by obtaining the competence of the set $Tr(P) \backslash Ck(S)$ (Yu and Zhang, 1990).

Between these sets of competencies takes place the following relationship:

$$Ck(S) \subseteq Tr(P) \subseteq HD$$

For a specific problem/ project, the *common HD* can be determined, which can be treated as a space containing all the competencies related to the project and which can be used in the analysis of the project solution efficiency by different participants of the team with different competencies (Yu, 1990; Yu and Zhang, 1990).

Analysis of competence initiated by (Yu and Zhang, 1990) addresses several issues related to the nature of human capacity:

- Extension of the competence is always associated with some effort and time. Therefore, one can control the functions of cost and time involved.
- Competence can be extended in many ways. Effective extension of competence should minimise the associated cost and time.
- The new competence can be easily achieved if it is coincident with the knowledge and skills represented by the current set of competencies. Therefore, in the analysis of competence development, the relations existing between competencies have to be taken into account.

Skills included in the set of competencies are inter-related. For any pair of competence *a* and *b* in *HD* space, if competence *b* can be achieved with competence *a* in a finite time then there is relationship for them. In this case, competence *a* is defined as a *base competence* for competence *b*. Therefore with existing relations between the competencies, the problem of domain space can be mapped by a diagraph (Yu and Zhang, 1990; Li, 1999).

Knowledge and skills needed to acquire new *group* competencies can be accumulated from many different sources. Therefore the more *a team of a project* has different *base individual competencies*, the faster the needed group competence will be acquired. This is due to a synergetic effect of combined individual competencies, which is always stronger than the impact of individual base competencies. Some competence models take into account this phenomenon and introduce complex competencies to include the strongest base relationship for several competencies considered together compared to the forces of relationships considered separately. In these models complex competence are reflected by distinct nodes in graphs depicting areas of specific sets of competencies (Yu and Zhang, 1990).

It is assumed that any competence *b* can be achieved from competence *a* if there are no time constraints. Therefore, it is possible to define the function f which can be described as the Cartesian product $HD \times HD$, where for each pair of competence is assigned the value of the set of real numbers and having the following characteristics (Yu and Zhang, 1990):

$$(*)f(a,b) \geq 0, \quad f(a,b) = 0, \quad \text{if } a = b \tag{6.1a}$$

$$(**)f(a,c) \leq f(a,b) + f(b,c) \quad \forall a,b,c \in HD \tag{6.1b}$$

Like the function $f(a,b)$ defining the temporal cost of achieving competence for the space HD, the functions associated with any other cost can be also defined. Most studies on the collection of competence relates to methods for the effective expansion of the competence with the criteria of time and cost (Yu and Zhang, 1990).

6.1.3 Representation of Competence as a Fuzzy Set

(Wang and Wang, 1998) found that the representation of competence as a classic set is insufficient because it does not provide information about the degree to master the competencies but only reflects the fact of the presence or lack of competence. The *fuzzy sets theory* developed by Zadeh (1965) was applied for the theory of competence. With the concept of a belonging function of elements of a set, it became possible to reflect the real nature of competence. In fact, besides finding that a person has a particular competence, you can quantify the extent to which the related skill has been mastered, qualitatively using linguistic values, for example. 'wrong', 'good' 'very good'. Moreover, from the point of view of competence requirements to solve a problem, a given competence may be required to some extent – not only fully 100%.

A fuzzy set is defined as follows (Zadeh, 1965):

$$A = \{(x, \mu_A(x)) | x \in X\}, \tag{6.2}$$

where $\mu_A(x)$ is the belonging function of element x of set A and which assigns each element of the X value in the range $[0;1]$, $\mu_A : X \rightarrow [0;1]$.

In the literature there are a number of studies describing the approach for competence modelling using fuzzy sets. Next, an overview of the most important terms based on the work of (Wang and Wang, 1998) is represented.

6.1.4 The Power of Competence

For each competence c, its power α is a function of the person having it project P or problem E, in the context of which it is considered.

$$\alpha : \{P\} \rightarrow [0;1] \tag{6.3}$$

So, competence can be measured by $c^{\alpha}(P)$. The competence for which $\alpha = 0$ is called *pseudo competence*.

The teaching experience shows that mastering any course is easier if you first master the fundamentals. Therefore it can be said that a certain course A may be the basis of another course B. Therefore the rate at which we are able to master course B depends on what degree we mastered the basic course A and how much these courses are dependent on each other. Based on these observations, one can define the concept of the relationship between the competencies.

6.1.5 Relationship and Basic Competence

If competence c_1 is a base competence of competence c_2, then between them exists a base relationship $r(c_1, c_2)$ of value $0 < r(c_1, c_2) = r_{21} \leq 1$. To simplify the relationship between c_1 and c_2 we denote it as $c_1 \xrightarrow{r_{21}} c_2$.

When $c_{21} = 0$, competence c1 is not the base competence of competence c2 and the relationship is noted $c_1 \mapsto c_2$. Competencies c_1 and c_2 are independent when $r_{12} = 0$ and $r_{21} = 0$.

6.1.6 Potential for Competence

The value of the relationship between competence and base competencies strength has an impact on the degree of acquiring new competencies. Therefore it can be concluded that the higher the value of the relationship base and the power of base competencies, the higher degree of obtaining a new competence. Thus: For any person of a team $s \in S$, if c1 about the power of α_1 is the only base competence for competence c_2, then the potential for this competence is $\beta_2 = \alpha_1 \cdot r_{21}$.

If a competence has a lot of base competencies of various powers then the process of this competence acquisition is done according to the *principle of maximum support*:

Assuming that $c_1^{\alpha_1}(s), \cdots, c_j^{\alpha_1}(s), \cdots, c_n^{\alpha_n}(s)$ are the base competencies c_i and that they have a relationship $r_{i1}, \cdots, r_{ij}, \cdots r_{in}$, then for person s the potential for acquisition of competencies c_i is given by:

$$\beta_j = \max_{j=1}^{n}(\alpha_j \cdot r_{ij}) \tag{6.4}$$

6.1.7 The Critical Level of Potential for Competence

Consider now the case when person $s \in S$ meets with problem/issue e which requires a competence c at a certain level α. If the person hasn't this competence, i.e. $\alpha = 0$, but has the potential to obtain this competence of a certain value γ, then he/she can easily acquire this competence at level α. Otherwise, the person will have to bear the extra effort associated with obtaining the same power competence c. The additional effort to obtain competence is the cost and is denoted by d.

For person $s \in S$ aimed at the solution of problem e, which requires competence c with power α (denoted $c^{\alpha}(e)$), the *critical level of potential* β to obtain this competence is a value in the range $[0;1]$ that

if $\beta \geq \gamma$ then $d = 0$, otherwise $d > 0$. $\tag{6.5}$

The value of the critical level γ depends on the acquired competence and is denoted by $\gamma(c)$. A person aiming to obtain competence c must have such a potential of base competencies that $\beta \geq \gamma(c)$. Such competence becomes an *owned (skill) competence*. In case $\beta < \gamma(c)$ then $\alpha = 0$ and competence c is called a **not-owned (non-skill) competence**. A set of owned competencies is denoted by $Sk(s)$ and a set of not owned competencies as $NSk(s)$.

Fuzzy complex competencies   Since the knowledge and skills needed to acquire a new competence can accumulate from various sources, the more one has different base competencies of some competence 'g', the faster one accumulates this competence. This is due to a synergetic effect of complex combined competencies, which is always stronger than the impact of individual base competencies.

If $c_1^{\alpha_1}(s), \cdots, c_j^{\alpha_1}(s), \cdots, c_n^{\alpha_n}(s)$ are composed on the complex competence C^{α} of the team S, then

$$C^{\alpha}(S) = \{c_1 \oplus \cdots c_j \oplus \cdots c_n\}^{MIN\{\alpha_1,\cdots,\alpha_j,\cdots,\alpha_n\}} \qquad (6.6)$$

where $\oplus$ refers to an aggregation of competence $c_j, j = 1, 2, \cdots, n$, which are components of a complex competence $C^{\alpha}(S)$. Moreover, the base complex competence is always in a stronger relationship with a given competence rather than its individual competencies.

Types of competence   A person can solve problem e only if $\beta(s) \geq \beta(e) = \gamma(c)$. This means that competence c must be within the set of owned competencies $Ck(s)$ and the person can intensify the power of competence c to a level $\alpha(e)$ to solve problem e.

Competence $c^{\alpha(e),\beta(e)}$, which is required to solve problem e because of the mastery degree by person s, can be classified into three types:

1) If $\beta(s) \geq \beta(s)$ and $\alpha(s) \geq \alpha(e)$, competence $c^{\alpha,\beta}(e)$ is called the competence of type (1).

2) If $\beta(s) \geq \beta(e)$ and $\alpha(s) < \alpha(e)$, competence $c^{\alpha,\beta}(e)$ is called the competence of the type (2).

3) If $\beta(s) < \beta(e)$, competence $c^{\alpha,\beta}(e)$ is called the competence of the type (3).

Case (1) means that person s has the competence to solve the problem e. In case (2), person s can enhance the power of the existing competence from level $\alpha(s)$ to level $\alpha(e)$. In case (3), person s can't enhance competence c, because it does not belong to the set of owned competencies $Ck(s)$. He/she must first increase the base potential $\beta(s)$ and only then intensify the competence c.

Methods of analysis of the cost extension of the person and project team competence Because of different types of competence representation, the models using fuzzy sets require other methods of analysis of the cost extension competence. At present in the literature you cannot find many of these methods. Moreover, in the latest development, a trend can be seen presenting extension process competence as a multiple-criterion problem, e.g. the method proposed by Lin (Lin, 2006).

Wang and Wang presented (Wang and Wang, 1998) a well-described and verified method of analysis of the cost expansion of the fuzzy sets of competencies. This method uses the approach to solving the problem of the competence extension based on a two-step procedure. The first stage of the procedure called the initial phase of the extension is to strengthen the power of competence type (2) so as to transform it into competence of type (1). The second stage of substantial extension of the competence is modelled as an optimisation task, where the cost of increasing the potential of base competencies type (3) is minimised. It leads to crossing their thresholds and as a result also transforming the competence of type (1).

The presented method of creating teams for the consortium is a multiple-criterion method, and the cost of the extension of competence is one of the criteria used in decision-making. Using for analysis of cost extension of the competence an additional multiple-criterion method should complicate the general method and create major implementation problems. Therefore, it seems sufficient here to use only a one-criterion model of the competence extension. In case, however, when it is necessary to use different criteria of the cost extension of competence (e.g. financial and time), they can be treated as a different decision criteria and then be aggregated together with all other criteria under consideration of the decision-making problem of the creation of teams for the consortium.

In view of the above arguments it was decided to use the method proposed by Wang and Wang (Wang and Wang, 1998). The exact method of solving the problem of extension of the competencies is described in the cited publication. However, in order to facilitate the analysis of the examples in this study, it was decided to change the specific assumptions of the used analysis method of the cost extension of competence.

Features of analysis methods of the extension cost of fuzzy sets of competencies As defined above set of project competencies is a collection of required competencies. If a person on a project team has the competence or base potential that exceeds the critical level, then a collection of such competencies creates a set of competencies, which is the following fuzzy set:

$$\{c_k^{\alpha_k,\beta_k}\} = \{(c_k,(\alpha_k,\beta_k))|0 \le \alpha_k \le 1; \forall k\} \tag{6.7}$$

The HD of the competencies is the set defined as follows:

$$\{c_k^{\alpha_k,\beta_k}(s)|k = 1,2,\cdots,n\} = HD^{\alpha,\beta}(P) \tag{6.8}$$

where $c_k^{\alpha_k,\beta_k}(s)|$ is the single or complex competence.

On the basis of the principle of minimal support, the set of competencies of the project team is the following:

$$Ck^{x,y}(P) = \min_{s \in S}\{c_k^{\alpha_k,\beta_k}(s)|\beta_k \ge \gamma_k(c_k), \ k = 1,2,\cdots\} \tag{6.9}$$

where $\quad x_k \equiv \alpha_k, \quad k = 1\cdots n, \quad x = [\,]_{n\times1}^T, \quad y_k \equiv \beta_k, \quad k = 1\cdots n,$ $y = [y_1\cdots y_n]_{n\times1}^T$, and x and y are the sub-vectors of vectors α and β, respectively.

The set of competencies not owned by the project team is defined as follows:

$$NCk^{\delta,\xi}(P) \equiv HD^{\alpha,\beta}(P)/Ck^{x,y}(P) \tag{6.10}$$

where $\alpha^T = [x,\delta]^T$ and $\beta^T = [y,\xi]^T$.

For any project the *P*/problem *E* set of competencies required to solve it is called the set of competencies required to solve this project/problem (*truly needed competence set*). It is defined as follows:

$$Tr^{\psi,\omega}(P) = \{Tr_i^{\psi_i,\omega_i}(s)|i = 1,2,\cdots,l\}. \tag{6.11}$$

where $\xi = [\xi_i]_{l\times1}^T, \xi_i = \alpha_i(Tr_i(P))$ and $\omega = [\omega_i]_{l\times1}^T, \omega_i = \beta_i(Tr_i(P))$ are respectively vector of powers and base potential of the individual elements of the set.

Each of the required competencies $Tr_i^{\xi,\omega}(s)$ contained in a set of required competencies can be one of three types. Competencies type (1) included in the set $Tr_i^{\xi,\omega}(P)$ are competencies already owned by project team P and there is no need to increase their power in order to solve the project/problem P. When all the competencies $Tr_i^{\xi,\omega}(P)$ of a project team P are type (1), this means that team P is able to solve project P without further increasing its competencies.

In the case of competencies of type (2) it is necessary to increase the power of these competencies in order to reach the level necessary to solve project P. Thus competencies of type (2) are competencies that persons of team $\{s\} = S$ must acquire in the first place, and the principle of their acquiring is the principle of *least resistance* (Wang and Wang, 1998).

This principle is based on choosing competencies of type (2) to extend in such a way that first the competencies for which the cost of extension is the smallest are extended. Increasing the power of competence type (2) can result in an increase of the base potential of some competence type (3) and as a result their transformation to competence type (2). This process is repeated until there is no longer any power type (3) which could be converted to type (2) by increasing the power of the other competencies of type (2). This process is called the *initial phase of the expansion.*

Following the initial phase of expansion, the status of the competence of project team P is defined by a set $Ck^{x^0}(P)$. This set contains two types of competence:

$$Ck^{x^0}(P) = \{Ck_j | j = 1, 2, \cdots, q < n\} \cup \{Ck_{q+k} | k = 1, 2, \cdots, n - q\}$$

$$(6.12)$$

where elements $Ck^{x^0}(P)$ are individual competencies and the elements $Ck_{q+k}^{x^0}(P)$ are the competencies composed of competence $Ck_j(P), j = 1, 2, \cdots, q.$

The competencies type (3), which after the initial phase of expansion are included in the set of not-having competencies, are denoted by $NCk^{\phi,\varphi}(P)|_E$. It is a subset of the initial set of competencies not possessed by project team P. Between these sets take place the following relationships:

$$NCk^{\phi,\varphi}(P)|_E \subset NCk(P) = HD(P)\backslash Ck(P) \qquad (6.13)$$

After the preliminary phase of the expansion of the competence, all the other competencies in the set $NCk^{\phi,\varphi}(P)|_E$ must be enhanced to the appropriate level required to solve the problem of E.

The set $Ck^x(P) = \{Ck_j^{x_j}(P)|j = 1, 2, \cdots, n\}$ is called a potential set of competencies $NCk^{\phi,\varphi}(P)|_E$ if and only if for all $i = 1, 2, \cdots, m$ there exists such j that $x_j \cdot r_{ij} \geq \varphi$, where r_{ij} is the relation between $Sk_j(P)$ and $NSk_i(P)|_E$.

So, having a potential set of competencies $Ck(P)$, the set $NCk^{\phi,\varphi}(P)|_E$ can be found without incurring any cost expansion of competence.

6.2 Team Collaboration Model and Method of Analysis of the Cost Extending Competence

6.2.1 Methods of Analysis of the Cost of Expanding Set of Competencies

The question of how most effectively to acquire a suitable set of competencies in order to effectively solve a specific problem is one of the most important questions among the issues related to the decision-making and analysis of competence sets (Feng and Yu, 1998).

For example, to obtain a title or certificate somebody must take a number of courses and practice. Each course can theoretically be represented by the skills that are elements of a competence set. Having skills g_i, the mastering of other skills g_j. takes a certain period of time and a certain expense. Then the question arises in which order should be acquired the individual skills needed for a certain title/certificate be acquired so that the total cost is the lowest (Feng and Yu, 1998).

Formally, the *process of expanding of the competence* set is meant as the way

$$\Gamma = (g_k, g_2, \cdots, g_n),$$

stretched on the acycled graph, defined by a set of competencies

$$Tr(E) \backslash Sk(P) = (g_1, g_2, \cdots, g_n).$$

Studying the process of the expansion of competence should determine the cost of achieving the new competence from the current set $Sk(P)$. Assuming that c is the cost function defined on $HD \times HD$, it can be defined as a C function determining the cost of obtaining a new competence from any currently owned set of competencies (Yu and Zhang, 1990).

In the simplest case, the *optimal process of extension* is obtained by using the principle of selecting of such competence to set $Sk(P)$, which minimises the cost at a given stage of extension (Yu and Zhang, 1990):

$$C(Sk(P) \cup \{g_{k_1}, g_{k_2}, \cdots, g_{k_{i-1}}\}, g_{k_i})$$

The application of the above method is limited to the case where the cost function $c(g_i, g_j)$ is symmetrical (i.e. $c(g_i, g_j) = c(g_j, g_i)$ for each i and j). For the case, when the cost function is asymmetric (i.e. $c(g_i, g_j) \neq c(g_j, g_i)$), the digraph describing the relationship between competencies includes the cycles. Shi and Yu (Shi and Yu, 1999) proposed a method of using integer programming to find the optimal expansion process. This method also takes into account the existence of complex and indirect competencies. They can be the competencies unowned by a person, who solves a problem and the competencies unnecessary to solve the problem. However, they can significantly facilitate obtaining required competencies. On the other hand, Li and Yu (Li *et al.*, 2000) used the conception of deductive graphs to develop a method of integer programming for problem-solving expanding a set of competencies for the case when complex and indirect competencies are in a digraph that don't contain cycles.

This review of the methods of analysis of the competence-extending cost provides many effective and thoroughly explored methods that can be used in the proposed method of selecting teams for a project. However, in recent years, works introducing a new enhanced approach to modelling competence have appeared. Namely, the methods, using the fuzzy sets instead of the hitherto conventional sets, are developed (Wang and Wang, 1998).

Domain area of the project P

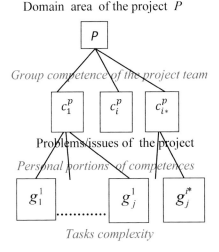

Figure 6.1 Hierarchical graph of the project domain.

6.2.2 Team Collaboration Model and Competence Expansion Algorithm to Perform the Project Task

Basic components of the project situation

a) $\{P, N^P, S^P\}$ – participants of the project situation, where:

N^P is the co-ordinator of project P,

$S^P = \{s_k\}$ is the project team, where $k = 1, 2, \cdots , k$ - index of team participant.

b) $\{p, C^P, G^P\}$ is the ontological/hierarchical graph of the project domain (Fig. 6.1), where:

p is the is a root vertex of the graph

$C^P = \{c_1^p, c_2^p, \cdots , c_{i*}^p\}$, $i = 1, \cdots , i^*$ are the competence portions of the project/subordinate nodes of the ontology graph,

$G^P = \{G(c_1^p), G(c_2^p), \cdots , G(c_{i*}^p)\}$ is a set of tasks, having to solve in the project,

$G(c_i^p) = \{g_1^i, g_2^i, \cdots , g_{j*}^i\}$, $j = 1, \cdots , j_i^*$ are subordinate tasks of the project competence c_i^p,

c) $\hat{G} = \cup_i G(c_i^p) = \{g_j^i\}$ is the set/repository of the tasks, where:

g_j^i is task $'j'$ consisting for competence portion $'i'$,

$j = 1, \cdots, j_i^*$ is the index of task

$i = 1, \cdots, i^*$ is the index of acquired competence

$c(g_j^i), j = 1, \cdots, j^*, i = 1, \cdots, i^*$ is the degree of complexity of a task, which can be expressed in numerical scale (depends on number of concepts included in the task, solution method, etc.)

$$0 \leq c(g_j^i) \leq 1$$

d) $R_g^q = r(q_i^p, g_i^j), j = 1, \cdots, j^*, i = 1, \cdots, i*$ are relationships between the vertices of the graph P.

If task g_j^i is a base task of problem q_i^p, then between them exists the *relationship* $r(q_i^p, g_j^i)$ of value $0 \leq r(q_i^p, g_j^i) \leq 1$.

e) Power of personal competence required for the solution of the problem

$$\alpha(c(q_i^p)) = \max_{j=1,\cdots,j^*} (c(g_j^i) r(q_i^p, g_j^i))$$

is determined on *the principle of maximum*.

f) Minimal potential of personal competence, required for the solution of the problem

$$\beta(c(q_i^p)), \ i = 1, \cdots, i^*$$

g) Relation of a project and problems

$$R_i^p = \{r(p, q_i^p)\}, \ i = 1, \cdots, i^*$$

If problem q_i^p is a base problem of project P, then between them exists the *relationship* $r(p, q_i^p)$ of value:

$$0 < r(p, q_i^p) \leq 1$$

h) Power of group/team competence required for the solution of the project

$$\alpha(c^p) = \min_{j=1,\cdots,j^*} (c(q_i^p) r(p, q_i^p))$$

is determined on *the principle of minimum*.

Decision variables   Project team $S^P = \{s_k\}$, where $k = 1, 2, \cdots, k^*$ is the index of the student/participant.

Numerical characteristics of the project participant:

$\alpha(c(s_k))$ – power of competence of participant $s_k \in S^P$

$\beta(c(s_k))$ – potential of competence of participant $s_k \in S^P$

Matrix of assignment of the project problems/issues to participants of project team

$$H = \|h(q_i^p, s_k)\|, \ i = 1, 2, \cdots, s^*,$$

where

$$h(q_i^p, s_k) = \left\{ \begin{array}{l} 1, \text{ if the problem } q_i^p \text{ is assigned to participants } s_k \\ 0, \text{ otherwise} \end{array} \right\}$$

Constraints on decision variables   A one-to-one relation of the project problems and team participants

$$\sum_{s \in S} h(q_i^p, s_k) = 1, \ \sum_{q_i \in Q^P} h(q_i^p, s_k) = 1$$

i.e. only one problem assigned to one participant and vice versa.

Criteria   Three kinds of individual competence:

1) The power of person competence more required competence of the problem

$$\alpha(c(s_k)) \geq \alpha(c(q_i^p)), \ s_k \in S_1, \ q_i \in Q_1$$

It is no important potential of person $\beta(s_k)$ and minimal potential of personal competence $\beta(c(q_i))$, required for the solution of the problem. Here no cost and time are required for solving the problem q_i.

2) The power of person competence less required competence of the problem

$$\alpha(c(s_k)) < \alpha(c(q_i^p))$$

and potential of the person $\beta(s_k)$ plus minimal potential of personal competence $\beta(c(q_i))$, required for the solution of the problem

$$\beta(s_k) > \beta(c(q_i))$$

Here, cost and time are required for solving problem q_i

$$f[\alpha(c(q_i^P)) - \alpha(c(s_k))], \ s_k \in S_2, \ q_i \in Q_2$$

3) The power of person competence less required competence of the problem/issue

$$\alpha(c(s_k)) < \alpha(c(q_i^P))$$

and potential of the person $\beta(s_k)$ less minimal potential of personal competence $\beta(c(q_i))$, required for the solution of the problem

$$\beta(s_k) < \beta(c(q_i))$$

Here additional efforts are required to increase the potential of the person, and after that some cost and time are required for solving the problem q_i

$$f[\alpha((s_k))], \ s_k \in S_3, \ q_i \in Q_3$$

Therefore the criterion is the following:

$$F = f_1 + f_2 + f_3 =$$

$$= \sum_{s_k \in S_1} \sum_{q_i \in Q_1} h(q_i^P, s_k) \times 0+$$

$$+ \sum_{s_k \in S_2} \sum_{q_i \in Q_2} h(q_i^P, s_k) f[\alpha(c(q_i^P)) - \alpha(c(s_k))]+$$

$$+ \sum_{s_k \in S_3} \sum_{q_i \in Q_3} h(q_i^P, s_k) f[\alpha(c(q_i^P)) - \alpha(c(s_k))] + g[\beta(c(q_i)) - \beta(s_k)]$$

Problem formulation For a team S solving project P, after the initial phase of the personal competence extension, the competencies of the project team are defined by a set $C_k^{x^0}(P)$ and the competencies of not having that project team in the context of problem E are defined by a set $NC_k^{x^0}(p)/_E$. Then the following relationship is real (Wang and Wang, 1998):

$$R \circ x^0 < \varphi$$

where $\circ$ is the operator *max – product*.

$R = [r_{ij}]_{m \times n}$ is the matrix of relationship between the elements of sets $C_k^{x^0}(P)$ and $NC_k^{x^0}(p)/_E$.

If $[0, 0, \cdots, 0]_{n \times 1}^T \le x^0 \le x \le [1, 1, \cdots, 1]_{n \times 1}^T$, then

$$R \circ x^0 \le R \circ x.$$

From the above rules of monotonic it results that increasing the personal competence power in set $C_k(P)$ from levels x^0 to x can increase the base potential of competence in set $NC_k(P)$. Therefore, based on the definition

$$\{c_k^{\alpha_k, \beta_k}\} = \{(c_k(\alpha_k, \beta_k)) | 0 \le \alpha_k \le 1; 0 \le \beta_k \le 1, \forall k\}$$

is possible to find such a vector x, $x^0 \le x \le [1, 1, \cdots, 1]_{n \times 1}^T$, for which the set $C_k(P)$ is potential set $NC_k(P)$ for the low cost of competence extension.

Therefore the model of competence expansion can be defined as follows:

$$\min \sum_{k=1}^{q<n} c_k(x_k) - c_k(x_k^0)$$

with restrictions:

$$R \circ x^0 \ge \varphi$$

$$[0, 0, \cdots, 0]_{n \times 1}^T \le x^0 \le x \le [1, 1, \cdots, 1]_{n \times 1}^T$$

$$x_{q+k} = \min_{j+1}^{q} \{x_j | j \in J_{q+k}\}, \forall k = 1, 2, \cdots, n-q$$

where R is a matrix of the relationship between the elements of sets $NC_k(P)/_E$ and $C_k(P)$ and c_k is a monotonic increasing function of cost. The last restriction is constructed using the relationship between simple and complex competence, and J_{q+k} is an index in which each iteration refers to the component of the complex competence.

6.2.3 Case Study

As a case study let's examine the project of forming an ontological model of a relational database described in Chapter 4.4.

In Table 6.1 the list of competencies relevant for each compe-tence project tasks is represented.

Here are used notations, adopted in the team collaboration model (Chapter 6.2.2).

1. *Basic components of the project situation*
   Personal competence of task solution
   Problems and Competencies portions
   $$\{e_1^P, c_1\}, \{e_2^P, c_2\}, \{e_3^P, c_3\}$$
   Indexes of task
   $$G_1 = \{g_{11}\ g_{12}\ g_{13}\ g_{14}\},$$
   $$G2 = \{g_{21}\ g_{22}\ g_{23}\ g_{24}\ g_{25}\ g_{26}\},$$
   $$G3 = \{g_{31}\ g_{32}\ g_{33}\ g_{34}\}$$
   Complexity of tasks
   $q(g_{11}) = 0.6; q(g_{12}) = 0.5; q(g_{13}) = 0.4; q(g_{14}) = 0.5;$
   $q(g_{21}) = 0.8; q(g_{22}) = 0.7;\quad q(g_{23}) = 0.6;\quad q(g_{24}) = 0.4;$
   $q(g_{25}) = 0.5; q(g_{26}) = 0.7;$
   $q(g_{31}) = 0.6; q(g_{32}) = 0.3; q(g_{33}) = 0.6; q(g_{34}) = 0.7;$
   Relationship values task-competence portion
   $R(c_1, g_{11}) = 0.3;\quad R(c_1, g_{12}) = 0.3;\quad R(c_1, g_{13}) = 0.1;$
   $R(c_1, g_{14}) = 0.2;$
   $R(c_2, g_{21}) = 0.8;\quad R(c_2, g_{22}) = 0.6;\quad R(c_2, g_{23}) = 0.4;$
   $R(c_2, g_{24}) = 0.3; R(c_2, g_{25}) = 0.3; R(c_2, g_{26}) = 0.7;$
   $R(c_3, g_{31}) = 0.8;\quad R(c_3, g_{32}) = 0.3;\quad R(c_3, g_{33}) = 0.6;$
   $R(c_3, g_{34}) = 0.7;$
   Power of personal competence, required for solution of task
   $q(g_{11}) \cdot R(c_1, g_{11}) = 0.18;\qquad q(g_{12}) \cdot R(c_1, g_{12}) = 0.15;$
   $q(g_{13}) \cdot R(c_1, g_{13}) = 0.04; q(g_{14}) \cdot R(c_1, g_{14}) = 0.1;$
   $q(g_{21}) \cdot R(c_2, g_{21}) = 0.64;\qquad q(g_{22}) \cdot R(c_2, g_{22}) = 0.42;$
   $q(g_{23}) \cdot R(c_2, g_{23}) = 0.24; q(g_{24} \cdot R(c_2, g_{24}) = 0.12;$
   $q(g_{25}) \cdot R(c_2, g_{25}) = 0.15; q(g_{26}) \cdot R(c_2, g_{26}) = 0.49;$
   $q(g_{31}) \cdot R(c_3, g_{31}) = 0.48;\qquad q(g_{32}) \cdot R(c_3, g_{32}) = 0.09;$
   $q(g_{33}) \cdot R(c_3, g_{33}) = 0.36; q(g_{34}) \cdot R(c_3, g_{34}) = 0.49;$

Table 6.1 Structure of the project 'Develop a relational database model in the given subject area'

Tasks for a specialist in the domain area	Tasks for system analyst	Tasks for a specialist in information technology and software engineering
1. Select the subject area and give a description of the domain (e.g. production company).	1. Prove that the structure of each information object meets the requirements of the third normal form.	1. Formulate and describe typical inquiries in the domain area and determine their relevant functional link.
2. Specify the information objects of the domain.	2. Determine the matrix of relationships between objects and build a graph of relationships.	2. Use rules for processing complex function links, which implements typical queries to form simple links.
3. Identify sets of key and non-key attributes of each information object.	3. To analyse and eliminate relationships of type M : M, replace them into relationships of type $1 : M$ by introducing new natural or artificial information objects.	3. Specify the parameters of the physical organisation of the database and implement a physical structure in a Microsoft Access environment.
4. Specify the characteristics of each object and the attribute of the object in tabular form.	4. Identify key and non-key attributes of new objects. Provide a modified matrix and a graph of relationships.	4. Formulate common queries in the data manipulation language (SQL type).
	5. Describe the semantics and characteristics of relationships in tabular form.	
	6. Develop a canonical structure of the relational model of the domain using the hierarchical graphing algorithm.	
	7. Build a relational database schema and describe the logical database structure in a data description language (DDL).	

Table 6.2 Power of group competence, required for the solution of the problem

Number of variables $k = 1, \cdots, 6$	Assigned problem (s_i^p, e_j^p)	Vector of personal competence to solve the problem $A(c_k) = (\alpha_1, \alpha_2\alpha_3) = R(p, e_j^p) \times \alpha(c(s_i^p, e_j^p)),$ $i = 1, 2, 3; j = 1, 2, 3$	Group competence to solve the project $A(c_k) = Min(\alpha_1, \alpha_2\alpha_3)$
1	< 11, 22, 33 >	< 0.24;0.64;0.48 >	0.24
2	< 12, 21, 33 >	< 0.4;0.15;0.48 >	0.15
3	< 13, 22, 31 >	< 0.24;0.64;0.12 >	0.12
4	< 11, 23, 32 >	< 0.24;0.18;0.32 >	0.18
5	< 12, 23, 31 >	< 0.4;0.18;0.24 >	0.18
6	< 13, 21, 32 >	< 0.24;0.5;0.32 >	0.24

2. *Group/Team competence required to solve the problem*
   Power of competence, required for the solution of the problem

$$\alpha(c_1, e_1) = \max_{j=1\cdots4}(q(g_j^1) \times R(c_1, g_j^1)) = 0.18;$$

$$\alpha(c_2, e_2) = \max_{j=1\cdots6}(q(g_j^2) \times R(c_2, g_j^2)) = 0.64;$$

$$\alpha(c_3, e_3) = \max_{j=1\cdots4}(q(g_j^3) \times R(c_3, g_j^3)) = 0.49;$$

Minimal potential of group competence, required for the solution of the problem

$\beta(c_1(e_1)) = 0.4;$
$\beta(c_2(e_2)) = 0.5;$
$\beta(c_3(e_3)) = 0.4;$

Relationship value project-competence portion

$R(p, c_1^P) = 0.3;$
$R(p, c_2^P) = 0.8;$
$R(p, c_3^P) = 0.6;$

Power of group competence, required for the solution of the project

$$\gamma(c^P) = \min_i(\alpha(c(e_i^P)) \times R(p, c_i^P)) =$$

$$=min(0.144, 0.512, 0.294) = 0.144$$

3. *Characteristics of project team participations (students)*
   Team's participant index

s_1^P, s_2^P, s_3^P

Power of competence of the team's participant

$$\alpha(c(s_1^P)) = \{\alpha(c(e_1^1)) = 0.8), \alpha(c(e_2^1)) = 0.5), \alpha(c(e_3^1)) = 0.4)\}$$

$$\alpha(c(s_2^P)) = \{\alpha(c(e_1^2)) = 0.5), \alpha(c(e_2^2)) = 0.6), \alpha(c(e_3^2)) = 0.3)\}$$

$$\alpha(c(s_3^P)) = \{\alpha(c(e_1^3)) = 0.3), \alpha(c(e_2^3)) = 0.4), \alpha(c(e_3^3)) = 0.8)\}$$

Potential of competence of team's participant

$\beta(c(s_1^P)) = 0.3$
$\beta(c(s_2^P)) = 0.6$
$\beta(c(s_3^P)) = 0.3$

4. *Determining of decision variables*

Power of required personal competence to solve the problem

$$\alpha(s_1, 1) = 0.8, \alpha(s_1, 2) = 0.5, \alpha(s_1, 3) = 0.4$$
$$\alpha(s_2, 1) = 0.5, \alpha(s_2, 2) = 0.8, \alpha(s_2, 3) = 0.4$$
$$\alpha(s_3, 1) = 0.3, \alpha(s_3, 2) = 0.4, \alpha(s_3, 3) = 0.8$$

5. Optimal problem assignment to solve the project (s_i^p, e_j^p) (Table 6.2)

$$K : A(c_k) = Max\{Min(\alpha_1, \alpha_2, \alpha_3)\} \longrightarrow k = 1, < 11, 22, 33 >$$

Power of possessed personal competence to solve the problem

$$\alpha(c(s_1)) = R(p, e_1^p) \cdot \alpha(c(e_1^1) = 0.3 \cdot 0.8 = 0.24;$$
$$\alpha(c(s_2)) = R(p, e_2^p) \cdot \alpha(c(e_2^2) = 0.8 \cdot 0.8 = 0.64;$$
$$\alpha(c(s_3)) = R(p, e_3^p) \cdot \alpha(c(e_3^3) = 0.6 \cdot 0.8 = 0.48;$$

Comparison of required and possessed power of group/team competence

$$\alpha(c_1^P) = 0.18 < \alpha(c(s_1^P)) = 0.24,$$
$$\alpha(c_2^P) = 0.64 > \alpha(c(s_2^P)) = 0.48,$$
$$\alpha(c_3^P) = 0.49 > \alpha(c(s_3^P)) = 0.48.$$

Comparison of required and real potential of group/team competence

$$\beta(c_1^P) = 0.4 > \beta(c(s_1^P)) = 0.3,$$
$$\beta(c_2^P) = 0.5 < \beta(c(s_2^P)) = 0.6,$$
$$\beta(c_3^P) = 0.4 > \beta(c(s_3^P)) = 0.3.$$

6. *Determining of competence expansion*

$$F = f_1 + f_2 + f_3 =$$

$$= \sum_{s_k \in S_1} \sum_{e_i \in E_1} h(e_i^p, s_k) \cdot 0 +$$

$$+ \sum_{s_k \in S_2} \sum_{e_i \in E_2} h(e_i^p, s_k) f[\alpha(c(e_i^p)) - \alpha(c(s_k))] +$$

$$+ \sum_{s_k \in S_3} \sum_{e_i \in E_3} h(e_i^p, s_k) f[\alpha(c(e_i^p)) - \alpha(c(s_k))] + g[\beta(c(e_i)) - \beta(s_k)]$$

$$= 0 + (0.64 - 0.48)w_1 + (0.49 - 0.48)w_1 + (0.4 - 0.3)w_2 =$$

$$= 0.16 \cdot 10 + 0.1 \cdot 100 = 11.6,$$

where w_1 and w_2 are weighting factors corresponding to increase the power of competence c_2 for participant s_2 and potential of competence c_3 for participant s_3.

6.2.4 Summary

1. To summarise the above considerations, competence is a general concept, which defines the ability to perform different patterns of behaviour based on accumulated knowledge and experience, while the qualifications relate to all kinds of formal evidence confirming possessing by the person of the specific knowledge and skills. Simply put, the qualification is formal evidence of specific competencies.

2. Computer-aided management of human resources in the project requires the use of a formal model of competence, which enables to quantify the usefulness of the research team to participate in the project. The above definitions and the integrated model of the competence reflect only the nature of the competence and don't provide tools for quantitative analysis of competence. In cases when exact quantitative analysis of competence is required, it is necessary to rely on a model that will provide mathematical foundations and tools to carry it out. This model can precisely describe the competencies, their comparison, determination of the cost of the competence increase, determination of the adequacy of the competence of the individual to the aim of the tasks and solving of many other problems of a quantitative nature.

   In the case study, a practical method of the team's knowledge estimation in knowledge and innovative organisations is proposed.

3. The results can be used by a decision maker or on the basis of the organisation consisting of several (OUs). On the basis of knowledge assessments of participants of a project or OUs, the method allows the estimatation of the cost of a team's training necessary to meet the requirements of the project. The competence expansion costs are then used as criteria to assign project tasks to participants of the project defined. All steps of the proposed approach were illustrated in the case study.

6.3 Cost Estimation Algorithm and Decision-Making Model for Curriculum Modification in Educational Organisation

The decision about curriculum modification usually takes place at the knowledge level, mainly with consideration of individual academic staff competencies and qualifications. However, traditional approaches to cost estimation of curriculum modification are focused on material resources only. In the paper (Kushtina *et al.*, 2009) the authors present a cost estimation method and decision model for curriculum modification in educational organisations. The proposed method works at the knowledge level and employs competence sets as knowledge representation models in educational organisations. Authors used the theory of hierarchical, multi-level systems in order to define the model of the decision-making process of curriculum modification and its dimension. On the basis of this and also the use of a fuzzy competence model, the cost estimation algorithm in the form of a group competence expansion algorithm (GCEA) is proposed. The algorithm focuses on the cost of staff competence expansion caused by the knowledge development process.

The problem of cost and resource estimation in education is a common research issue in operational research (Fandel and Gal, 2001; Kwak and Lee, 1998). However, so far the analysis was focused on the material aspect of education. In the incoming knowledge era the methods of knowledge resource estimation are becoming more and more important. The new economy requires effective methods for assessing the level of knowledge resource utilisation (Freeman, 2001). Problems related to knowledge management exist especially in the university environment. In their research work and projects several authors (e.g. "Bologna Working Group on Qualifications Frameworks" (2005); Sanchez (2004)) came to the conclusion that knowledge management in an educational organisation should be based on competencies and qualifications. The methods of competence model theory (Yu and Zhang, 1990) can play a major role in the cost estimation algorithm due to their ability to assess cost and resources required for students and academic staff to develop certain knowledge in a specific domain.

Modern universities can be seen as continuously changing learning organisations (Johnes, 1996). One of the important factors is academic staff development, which allows universities to actively participate in scientific research and grant acquisition processes and at the same time to propose a curriculum attractive for the market (Alexander, 2000). A university's ability of dynamic curriculum modification or redesign is crucial for surviving on the open educational market, forced by the changing educational standards, the student's expectation and the labour market (Saunders and Machell, 2000; Tariq *et al.*, 2004). This situation is especially noticeable in the IT market(Byrne and Moore, 1997; Prabhakar *et al.*, 2005).

Educational institutions have to adapt their curriculum to the continuously changing IT sector (Hilburn *et al.*, 1999). Such process is always related to staff competence update (usually expansion) due to new technology, innovative theories or new application of existing ones. In their article Kushtina et al. (2009) propose a cost estimation method of curriculum modification. The proposed approach takes advantage of the theory of hierarchical, multi-level systems in order to define the decision-making process of curriculum modification. Then, on the basis of the fuzzy competence model the cost estimation algorithm in the form of a GCEA is proposed. The algorithm focuses on the cost of staff competence expansion caused by the knowledge development process. The rest of the chapter is organised in the following way: in the next part the decision model of curriculum modification is proposed. In order to create such a model, system analysis of a typical educational organisation is performed, taking advantage of the hierarchical, multi-level systems theory (Mesarovic *et al.*, 1970). Section 6.6 presents the proposition of the cost estimation method for curriculum modification in an educational organisation, which is based on fuzzy competence set theory (Yu and Zhang, 1990). Section 6.7 demonstrates the discussed issues on an example.

6.4 Conceptualisation of the Curriculum Modification Decision Model

Problem description   An educational organisation is considered as a complex system aimed at delivering knowledge to students in order to expand their competencies to the expected level (Tavares, 1995).

There are many approaches to describing each of the aspects of an educational organisation's functioning:

- Formal description of the organisation's structures (organisation theory)
- Description of the behaviour and motivation of the participants of the processes taking place in organisations (games theory)
- Description of the organisation's activity through modelling of basic functions (system theory)

Each of the mentioned approaches enables a deep analysis of the organisation's activity; however, none of them considers all of the processes occurring in the organisation and influencing its life cycle at the same time. System thinking makes it possible to distinguish the main processes, to describe their interactions and to find a methodological approach to creating an organisation management system that ensures the best-possible conditions for the organisation' functioning.

As a theoretical and methodological base for educational organisation analysis we use the approach developed in the theory of hierarchical multi-level systems (Mesarovic *et al.*, 1970). In that work, the authors introduce the idea of a multi-level hierarchical organisation structure that considers different functioning aspects of the organisation. The hierarchical multi-level analysis is popular in the literature, e.g. modelling social systems (Miklashevich and Barkaline, 2005), and multidimensional complex software systems (Gómez *et al.*, 2001). System analysis, according to this approach, considers the hierarchical nature of organisation management in two dimensions: (i) distinguishing the layers of decomposition of the problems the organisation is facing and (ii) deciding the order of decisions in the decision-making process when solving problems.

6.5 Functional Schema of Educational Organisation

As a result of system analysis the functional schema of an educational organisation is formulated (Fig. 6.2).

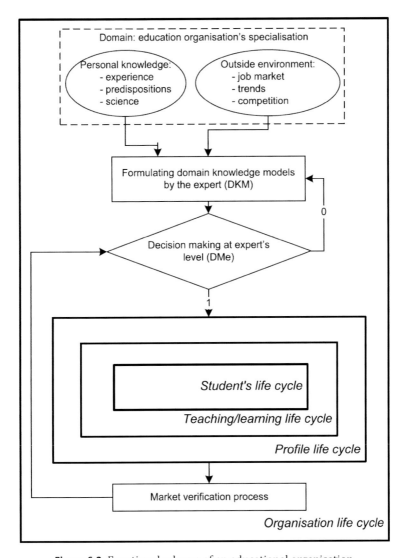

Figure 6.2 Functional schema of an educational organisation.

The functional schema can be described as a process of sequential knowledge processing during:

- Syllabus preparation
- Providing education services

- Developing didactic materials
- The process of acquiring competencies based on a specified knowledge model
- Statistical evaluation of students' progress

On the one hand, the functional schema recognises decision-making modules and decides the order of decisions in the decision-making process when solving problems. On the other hand, the functional schema helps distinguish the layers (cycles) of problems decomposition the organisation is facing. The working time of every cycle is deduced from existing standards, guidelines or procedures. During the cycle's working time, the management action is executed. The person who holds an adequate role is responsible for the management action's execution. The management action is executed in a proper manner if complete information is delivered to the right person. Because of the time horizon inter-relations and management actions characteristics, the cycles are organised as inbuilt, nested cycles (Fig. 6.2).

Figure 6.2 shows that the functional schema consists of four inbuilt management cycles. Special attention is dedicated to the organisation life cycle because the decision about curriculum modification is made at this level. The organisation life cycle consists of the profile life cycle, teaching/learning life cycle and student's life cycle (Table 6.3).

Management outlines   Considering management outline as a management model allows us to specify the decision-making process related to curriculum modification or redesign. The decision maker of level *i* can be described with the 6 tuple C_i:

$$C_i = \{MA_i, IV_i, F_i, PR_i, DF_i, MC_i\},$$

where:

> MA_i is the management activity coming from the top level,
> IV_i are the constraints (interfering variables) coming from the exterior environment,
> F_i is the feedback coming from the control process,

Table 6.3 The organisation life cycle

Name	Description	Decision-maker	Management process	Knowledge model
Organisation life cycle	Subsystem of educational organisation strategic management aims at maintaining a high position of organisation's graduates at the job market	Expert	Organisation management process	Domain knowledge
Profile life cycle	Subsystem of managing the process of adapting competence to the student's profile	Methodologist	Profile management process	Specialisation profile model
Teaching/learning life cycle	Subsystem meant to provide an intelligent and network space for the learning/teaching system, assuring effective use of network environment and developing or adapting knowledge repository to student's profile and students contingent	Knowledge engineer	Teaching/learning management process	Didactic materials model
Student's life cycle	Subsystem that allows maintaining and monitoring administrative correctness of the learning process, and evaluating the competence-gaining process with the given knowledge model and teaching/learning system	Teacher	Student management process	Learning process model

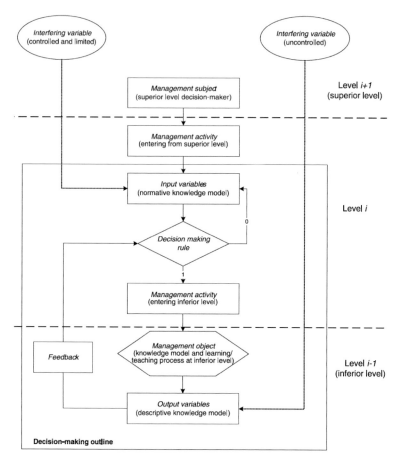

Figure 6.3 Management outline structure.

PR_i is the production rule, according to which the management activity for the bottom level is formulated,

$DF_i = MA_i + 1$ is the decision function, according to which the content of the management action entering the inferior level is estimated,

MC_i is the management cycle at level i, representing the time interval limiting the management action MA_i.

The diagram shown in Fig. 6.3 can be considered as a management outline (MO) with feedback Fi. The superior level decision maker $(i + 1)$ is the subject of management. The inferior level decision maker $(i − 1)$ together with the learning process are the objects of management. The central system-making element of an MO is the decision-making circuit that compares the normative knowledge (NK) model with the descriptive knowledge (DK) one. *The normative knowledge model* represents valid and proper behaviour rules, while the descriptive knowledge model impartially and objectively describes reality (Broens and de Vries, 2003). The normative knowledge model is formulated by the system's decision maker, on the basis of the management activity entering from the superior level. The descriptive knowledge model is specified by the inferior level decision maker.

As shown in (Zaikin *et al.*, 2006), the knowledge representation model most suitable for the researched system is a hierarchical concept graph G^P. If methodology and algorithms from (Zaikin *et al.*, 2006) are applied, knowledge models NK and DK can be presented in the form of hierarchical graphs G^P_{NK} and G^P_{DK}. With this approach, decision rule PR_i, on the basis of which decisions are made, can be described with the Kronecker symbol:

$$PR_i = \begin{cases} 1, \text{ if } G^P_{NK} \supseteq G^P_{DK} \text{ and } DH_i \leq T_i \leq DH_{i+1} \\ 0, \text{ otherwise} \end{cases},$$

where:

G^P_{NK} is the hierarchical graph of normative knowledge model,
G^P_{DK} is the hierarchical graph of descriptive knowledge model,
DH_i is the decision time horizon at level i,
DH_{i+1} is the decision time horizon at inferior level $i + 1$.

Each production rule PR_i consists of two conditions:

a) Normative knowledge graph G^P_{NK} covers descriptive knowledge graph G^P_{DK}:

$$G^P_{NK} \supseteq G^P_{DK}$$

b) The Decision-making period T_i is longer than decision horizonDH_i and shorter than decision horizon DH_{i+1}:

$$Dh_i \leq T_i \leq DH_{i+1}.$$

As can be seen in Fig. 6.3, when both conditions are fulfilled the decision maker formulates the management action for the inferior level $(i-1)$ on the basis of the decision functionDF_i. In this case, the decision function DF_i changes the normative knowledge model NK_{i+1} of the inferior level. Should one of the described conditions not be fulfilled, the decision maker stays at level i to develop and modify his/her own normative knowledge model NK_i.

In the case of the organisation life cycle, the domain knowledge model (DKM) can be described with the following tuple:

$$DKM = \{R, Ac, Kd\},$$

where:

R are roles,

Ac are activities,

Kd is the domain knowledge.

DKM is formulated by an expert on the basis of arising market demands for new products, technologies and enterprise organisation forms and for establishing new roles and redefining tasks for the domain specialists. The expert focuses on the domain that the given organisation specialises in.

The subsystem of educational organisation strategic management aims at maintaining a high position of organisation's graduates in the job market.

The expert's decision model (DMe) has the following form:

$$DMe = \{Mae, IVe, Fe, Pre, DFe, LCe\},$$

where:

MAe is the management activity,

IVe is the market demand for the specialisation,

SEe is the periodical graduates control in order to estimate their satisfaction in reference to market needs,

DFe is the decision function (creating new specialisation, modifying an existing one),

LCe is the organisation life cycle,

PRe is the expert's production rule in the following form:

$$PR_i = \begin{cases} 1, \text{ if } G^P_{NKe} \supseteq G^P_{DKg} \text{ and } DH_e \leq T \leq DH_m \\ 0, \text{ otherwise} \end{cases},$$

where:

G_{NKe}^P is the hierarchical graph of normative (domain) knowledge at the expert's level,

G_{DKg}^P is the hierarchical graph of graduate's descriptive knowledge,

DH_e is the expert's decision horizon,

DH_m is the methodologist's decision horizon.

6.6 Model of the Decision Support System for Curriculum Modification

Problem description   In the previous chapter we answered the question of at what management level and for what reason the decision about curriculum modification is made. In this chapter, we propose a practical method, which can be considered as part of a decision support system for a university. Applying the method would support a decision maker in the process of introducing a new curriculum into the educational organisation. Related procedures can be found in the literature (e.g. Trigueiros *et al.* (2006); however, the solutions proposed there are based on the heuristic approach and require a great expert's contribution.

The proposed method is designed to handle a typical decision-making situation in the process of introducing a new curriculum, that is assignment of new courses specified by the new curriculum to organisational units (OU) of the educational organisation. The assignment should be performed taking into consideration the cost of training the OU staff required to begin teaching a new course. This cost can be estimated by comparing competence requirements defined for a course and current competencies of the OU.

The *training cost estimation* problem requires defining a certain knowledge representation model that would allow to credibly reflect competencies of personnel in the field of teaching a new course. Another important feature of such a model would be the capability to estimate the cost of competence expansion necessary to reach the required level.

Both of the these assumptions are fulfilled by the method of fuzzy competence sets presented by Wang (Wang and Wang, 1998).

This method is based on the concept of HDs and competence sets proposed by Yu and Zhang (Yu and Zhang, 1990). According to this concept competencies represented by knowledge, skills and information required to solve a problem exist and are inter-related in the space called Habitual Domain (Yu, 1991). Thus, for a person challenging the problem it is possible to define his/her competence set as well as the set of competencies required to successfully solve the problem. By comparing these two sets the cost of expanding a person's competencies to the required level can be estimated.

In his work, Wang (Wang and Wang, 1998) established that representing competence as a classical set is not adequate to the nature of this phenomenon. In reality, apart from determining that one has a certain competence, it is also necessary to establish the level of this competence. Thus, the fuzzy set theory presented by Zadeh (Zadeh, 1965) was naturally applied to the concept of competence sets.

In the context of the problem examined in this chapter, fuzzy competence sets will be used to reflect competence of academic staff to teach courses. Competencies will be presented at a high level of generalisation, because one element of the competence set will be reflecting competence to teach one specific course on a one-to-one basis.

The proposed method is presented in Fig. 6.4 CC2005 stands for Computing Curricula 2005, the guide to undergraduate degree programmes in computing (ACM and IEEE, 2005). The algorithm consists of three main processes:

- Assessment of current competencies: In this process all OUs considered for teaching new curriculum are analyzed in order to find their competence sets. The analysis is conducted on the basis of the OUs' portfolios, their didactic experience and courses currently taught.

- Competence requirement assessment: The set of required competencies is assessed trough the analysis of the introduced curriculum.

- Group competence expansion cost analysis: The most relevant process of the proposed method. on the basis of the requirements of the new curriculum and competencies sets of OUs, an

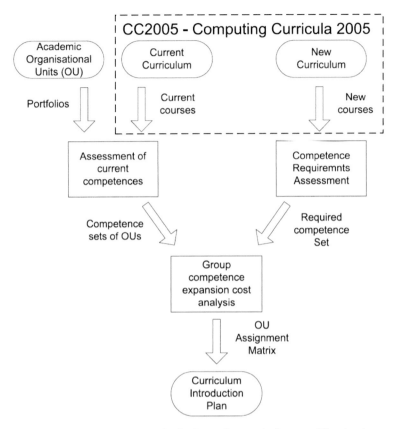

Figure 6.4 Cost estimation method schema for curriculum modification in an educational organisation.

optimal assignment of the OU to new courses has to be found. The criterion for the assignment is the minimal cost of OU staff training computed using competence expansion cost analysis.

6.6.1 Competence Representation Model

This section of the chapter contains a presentation of the fuzzy competence sets method elaborated on the basis of (Wang and Wang, 1998). Despite many existing articles on fuzzy competence sets, the author decided to include some basics of this concept in order to make the rest of the chapter clearer and to spare the reader from referring to external sources. In the early stage of the research on competence representation models, competence was presented as a classical set (Yu and Zhang, 1990) containing knowledge, skills and information necessary to solve a problem. However, assessing the presence of a competence in binary terms – one has a competence or not at all – turned out to be insufficient regarding the continuous nature of competence. On account of that, it was proposed to present human competence as a fuzzy set defined as follows (Zadeh, 1965):

$$A = \{(x, \mu_A(x)) | x \in X\},$$

where $\mu_A(x)$ is the membership function assessing the membership of an element x in relation to set A by mapping X into membership space $[0; \cdots, 1]$, $\mu_A : X - \{0\}$.

Regarding the definition of the fuzzy set it is possible to define the notion of fuzzy competence strength. For each competence g, its strength is a function of a person P or an event E in the context of which the competence is assessed:

$$\alpha : \{P \text{ or } E\} \rightarrow [0;1].$$

Thus, a competence can be described by $g^a(P)$ or $g^a(E)$. If competence strength equals zero, $a = 0$, then a competence is called pseudo competence. Research of the learning process shows that mastering a course is easier if the fundamental courses are learnt first (Ausubel et al., 1978). Thus, it is possible to state that a competence a can be a base of another competence. How fast and easy one can master a competence depends on 'how well' one has learnt the background competence and how close they relate to one another. On the basis of these remarks, it is possible to define the concept of a relation between competencies. If competence g_1 is a background competence of competence g_2, there exists a background relation denoted by $r(g_1, g_2)$ between these two competencies $0 < r(g_1, g_2) = r_{21} \leq 1$.

If $r_{21} = 0$ then competence g_2 is not a background competence of g_1. Competencies g_1 and g_2 are independent if $r_{12} = 0$ and $r_{21} = 0$. The value of the relation between two competencies and the strength of a background competence influence the speed and easiness of obtaining a new competence. Thus, the stronger is the background relation and strength of the background competence, the easier it is to obtain a new competence. For any person P, if competence g_1 with strength a_1 is the only background competence of g_2, then the background strength (potential) of obtaining g_2 equals $b_2 = a_1 \times r_{21}$.

If a competence has several background competencies then the learning process progresses according to the principle of maximal support. For person P, facing problem E requiring competence g with strength a (denoted by $g_a(E)$), the critical level c of the background strength b needed to obtain competence g is the value from range $[0;1]$, so that if $\beta \geq \gamma(g)$, then $c = 0$, otherwise $c > 0$. The value of the critical level c depends on the competence and is denoted by $c(g)$.

A person willing to obtain competence g has to possess such background strength of this competence, so that $\beta \geq \gamma(g)$. If the condition is fulfilled, then competence g is called the person's skill competence. Otherwise, g is called the person's non-skill competence. The set of all skill competencies is denoted by $Sk(P)$, while that of non-skill competencies is denoted by $NSk(P)$ (Wang and Wang, 1998).

According to Wang, competence $g^{a(E),b(E)}$, being the competence required to solve problem E, can be classified as one of the three following types:

- If $b(P) \geq b(E)$ and $a(P) \geq a(E)$, competence $g^{a,b}(E)$ is called person's P type (1) competence.
- If $b(P) \geq b(E)$ and $a(P) < a(E)$, competence $g^{a,b}(E)$ is called person's P type (2) competence.
- If $b(P) < b(E)$, competence $g^{a,b}(E)$ is called person's P type (3) competence.

In order to analyze the competence expansion process let us define the truly needed competence set $Tr^{\psi,\omega}(E)$ that contains all competencies necessary to solve the chosen problem E. Vector w defines required strengths of the competencies in Tr, and vector x defines their necessary background strengths. Briefly, the methods of optimal competence set expansion consist of determining the order of obtaining successive competencies that provides minimal cost (Chen, 2001; Feng and Yu, 1998; Li *et al.*, 2000). Competencies that need to be obtained are defined by the set $Tr^{\psi,\omega}(E)\backslash Sk^x(P)$, where x is the vector of strengths of competencies currently possessed by person P. This is often achieved through finding the shortest path in an oriented graph, in which vertices represent competencies and arcs represent the relations between them (Wang and Wang, 1998; Yu and Zhang, 1990).

6.6.2 Group Competence Expansion Algorithm

Educational organisations, even having highly skilled personnel, usually are not able to immediately introduce a new curriculum that differs from the one currently taught. Apart from administrative and infrastructure preparations, a new curriculum requires training the academic staff in new knowledge areas. At the same time, educational organisations try to avoid employing new and to optimally utilise the existing staff. Thus, there arise many questions regarding staff assignment and evaluation of the training cost. Competencies required from academic staff are very complex. Thus, a high level of competence generalisation was assumed for the purpose of this study. First of all, the method focuses on competencies of organisational structure units that are, or could be, responsible for teaching a certain domain of knowledge. The unit consists of one or several teachers and is the smallest entity in the organisational structure for which a didactic course can be allocated. Taking into consideration individual competencies of all teachers would not give any improvement to the model and would only increase its complexity.

The following assumptions can now be made:

Input data:
$P = \{P_i\}$, $i = 1, 2, \cdots, i^*$ – set of OUs of the educational organisation.
$HD^{\alpha,\beta} = \{g_k^{\alpha_k,\beta_k}|k = 1, 2, \cdots, \infty\}$ – space of all competencies related to teaching didactic courses.
R is the relation matrix of competencies in HD.
$Sk^x(p_i) = \{g_j^{x_j}|g_j^{x_j} \in HD^{\alpha,\beta}\}$ is the competence set of i-th organisational unit.
E is the problem to be solved.

$TR^{\psi,\omega}(E) = \{g_l^{\psi_i,\omega_i} | l = 1, 2, \cdots, l^*\}$ – set of competencies (courses) necessary to introduce a new curriculum.

$c(Sk^x(p_i), g_l^{\psi_i,\omega_i})$ – cost function of obtaining competence g_1 by unit P_i.

C_0 is the assumed maximum cost of introducing a new curriculum.

Control parameters:

$U = \|u_{il}\|$ is the allocation matrix.

$$U_{il} = \begin{cases} 1, & \text{OU } p_i \text{ teaches } g_i, \\ 0, & \text{otherwise.} \end{cases}$$

Assuming, that a course can be allocated only to a single unit:

$$\sum_{l=1}^{l^*} u_{il} = 1.$$

Criterion:

$$\Phi = \sum_{l=1}^{l^*} [u_{il} \cdot \delta_i \cdot c(Sk^x(P_i), g_l^{\psi_i,\omega_l})] = \min_U < C_0$$

It is assumed that all competence sets $Sk^x(p_i)$ and the truly needed competence set $Tr^{\psi,\omega}(E)$ are defined through an expert's analysis. The solution to the given task can be found using a heuristic algorithm (GCEA) proposed by the authors in (Kushtina *et al.*, 2009), and depicted in Fig. 6.5 In order to correctly interpret the algorithm it is necessary to define types of group competencies based on the three competence types presented earlier.

The problem that the authors examine is whether a group of educational OUs is capable of introducing a new curriculum. If all necessary competencies are covered by competencies of all units, then we can say that the organisation is able to introduce the new curriculum. Thus, it is necessary to present a new definition of competence types, one that takes into consideration joint competencies of a group. Hence, we can classify group competencies as follows:

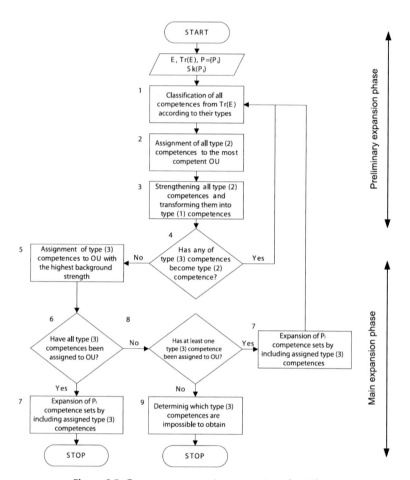

Figure 6.5 Group competencies expansion algorithm.

(1) If $\exists_{i=1}^{n}(\beta(P_i) \geq \beta(E) \wedge \alpha(P_i) \geq \alpha(E))$ then competence $g^{\alpha,\beta}(E)$ is called a type (1) group competence.

(2) If $g^{\alpha,\beta}(E)$ is not a type (1) group competence and $\exists_{i=1}^{n}(\beta(P_i) \geq \beta(E) \wedge \alpha(P_i) \geq \alpha(E))$, then this competence is a type (2) group competence.

(3) If $\forall_{i=1}^{n}\beta(P_i) < \beta(E)$ then competence $g^{\alpha,\beta}(E)$ is called a type (3) group competence.

The expansion process of group competencies has two phases: preliminary expansion phase and main expansion phase. The preliminary expansion phase contains activities that are performed to increase strengths of competencies already possessed by the group, when these strengths are too low in regard to the given problem E.

This corresponds to transforming all type (2) into type (1) competencies. The preliminary expansion phase contains steps 1–4 of the algorithm in Fig. 6.5 In the main expansion phase of the expansion process the group acquires new competencies that have not been skill competencies of any unit so far. It this phase, type (3) group competencies are obtained (steps 5–10 of the GCEA). In both expansion phases every competence from the set $Tr^{w}x(E)$ has to be allocated to exactly one unit on the minimal cost basis. In the preliminary expansion phase this is achieved by allocating a type (2) competence to the unit with the highest strength of this competence. In the main expansion phase, a type (3) competence is allocated to the unit with the highest background strength for this competence.

The main expansion phase of the expansion process strongly depends on the relations that exist between competencies. A new competence to teach didactic courses is obtained on the basis of its background competencies. However, it is possible that none of the units from set P have background competencies of type (3). This means that the competence cannot be obtained by any unit. In such case, problem E cannot be solved by team P. Such a situation requires outsourcing this course, what is associated with a high and difficult-to-estimate cost.

6.7 Case Study

This section describes a practical application of the GCEA presented in the previous section: introduction of an IT programme to a faculty which has only provided a computer science (CS) programme so far. In this case study we will use the experience acquired through working on a similar project done at the university and from the literature (Byrne and Moore, 1997; Kelting-Gibson, 2005).

Let's suppose the authorities of the faculty define a specific budget and time for the discussed project. This creates the problem of selecting academic staff capable of preparing the new programme within the assumed budget and time frame. CS and IT programmes are similar in certain domains, so some courses of CS can be directly adapted to IT and can be taught by the same staff. However, some courses have to be adjusted to IT requirements, e.g. their scope has to be broadened. Moreover, introducing a new programme requires several new specific courses. Therefore, the personnel responsible for the new programme must have or obtain competencies suitable for teaching new courses and to change the scope of courses that have been taught so far.

The faculty authorities need to verify whether the present staff is capable of preparing itself to teach the new discipline. Moreover, it has to be decided which university units will be responsible for which courses.

The university analysed in the example consists of five institutes that specialise in different domains. They are as follows:

P_1 – Institute of Computer Architecture and Telecommunication (ICAT)

P_2 – Institute of Computer Graphics and Multimedia Systems (ICGMS)

P_3 – Institute of Information Systems (IIS)

P_4 – Institute of Artificial Intelligence and Mathematical Methods (IAIMM)

P_5 – Institute of Software Engineering (ITP)

Each of the institutes has taught several courses specified in the CC2005 (ACM and IEEE, 2005) requirements for the CS programme. The courses had been assigned to them according to their competencies. However, each institute's competencies overlap and some units have competencies to teach courses not assigned to them. The competence sets of all institutes are presented in Table 6.4.

Basing on the Computing Curricula 2005 report (CC2005) (ACM and IEEE, 2005) it is possible to determine competencies necessary to introduce the IT programme. Topics included in the IT programme scope can be presented as competencies required from the academic staff. The strengths of these required competencies can be determined on the basis of weights defined for each course in the CC2005 report. A course weight has six possible values and is described as a discrete number from 0 to 5. For the weight to be useful for competence set analysis it has to be normalised

Table 6.4 Competence sets of all institutes

Institute	Courses assigned	Competence set Sk(P)
P_1	$g_4, g_5, g_6, g_7, g_8,$ $g_{28}, g_{29}, g_{30}, g_{31},$ $g_{32}, g_{33}, g_{34},$	$\{g_3, g_4, g_5, g_6, g_7, g_8, g_{14},$ $g_{19}, g_{28}, g_{29}, g_{30},$ $g_{31}, g_{32}, g_{33}, g_{34}\}$
P_2	g_{11}, g_{12}	$\{g_{11}, g_{12}\}$
P_3	$g_{14}, g_{15}, g_{17},$ $g_{19}, g_{20}, g_{22},$ $g_{23}, g_{24}, g_{25},$	$\{g_2, g_6, g_7, g_8, g_{11}, g_{14}, g_{15},$ $g_{17}, g_{19}, g_{20}, g_{22}, g_{23},$ $g_{24}, g_{25}, g_{31}, g_{33}, g_{34}\}$
P_4	g_{13}, g_{16}	$\{g_7, g_{13}, g_{16}\}$
P_5	$g_1, g_2, g_3, g_{10}, g_{26}, g_{27}$	$\{g_1, g_2, g_3, g_5, g_{10},$ $g_{14}, g_{15}, g_{19}, g_{22},$ $g_{26}, g_{27}, g_{32}\}$

into a continuous $[0;1]$ range. This normalisation can be done in the following manner: $\psi_i = 0.2...min(g_i)$, where ψ_i is the required strength of course competence and $min(g_i)$ is the minimal required course level according to the CC2005 report. Table 6.5 presents required competence strengths ψ, required competence background strengths x and existing competence strengths x for the five institutes under analysis.

Using the list of topics from Table 6.5 it is possible to determine the truly needed competence set for task E, which can be formulated as 'to introduce the IT programme into the university'. $Tr^{\psi,\omega}(E) = \{g_1; g_2; g_3; g_4; g_5; g_6; g_7; g_8; g_9; g_{11}; g_{14}; g_{15}; g_{17}; g_{18}; g_{19}; g_{22}; g_{23}; g_{24}; g_{25}; g_{26}; g_{27}; g_{29}; g_{30}; g_{31}; g_{32}; g_{33}; g_{34}; g_{35}; g_{36}\}$ On the basis of the competence sets of all institutes it is possible to determine group competence sets of the whole faculty:

Type (1) competencies: $g_1, g_3, g_4, g_5, g_{14}, g_{17}, g_{22}, g_{23}, g_{24}, g_{25}, g_{26},$ $g_{27}, g_{29}, g_{30}, g_{31}$

Type (2) competencies: $g_2, g_6, g_7, g_8, g_{11}, g_{15}, g_{19}, g_{32}, g_{33}$

Type (3) competencies: $g_9, g_{18}, g_{35}, g_{36}$

Beginning to teach a new programme requires preparing a set of new courses that have not been taught so far. Usually, it is necessary to strengthen or to obtain new competencies necessary for these

Table 6.5 List of competencies necessary to teach the IT program

Competence		IT min	IT max	ψ	ω	CS min	CS max	$x^{(1)}$	$x^{(2)}$	$x^{(3)}$	$x^{(4)}$	$x^{(5)}$
Programming fundamentals	g_1	2	4	0.4	0.78	4	5					0.8
Integrative programming	g_2	3	5	0.6	0.48	1	3			0.1		0.2
Algorithms and complexity	g_3	1	2	0.2	0.56	4	5				0.4	0.8
Computer architecture and organisation	g_4	1	2	0.2	0.56	2	4	0.3				
Operating systems principles and design	g_5	1	2	0.2	0.52	3	5	0.4				0.3
Operating systems configuration and use	g_6	3	5	0.6	0,6	2	4	0.6		0.1		
Net centric principles and design	g_7	3	4	0.6	0.54	2	4	0.4		0.2		
Net centric use and configuration	g_8	4	5	0.8	0.56	2	3	0.4		0.1		
Platform technologies	g_9	2	4	0.4	0.32	0	2	0.4				
Theory of programming languages	g_{10}	0	1			3	5					
Human–computer interaction	g_{11}	4	5	0.8	0.74	2	4		0.4	0.2		0.6
Graphics and visualisation	g_{12}	0	1			1	5		0.2			
Intelligent systems (AI)	g_{13}	0	0			2	5				0.4	
Information management (DB) theory	g_{14}	1	1	0.2	0.48	2	5	0.1		0.4		0.1
Information management (DB) practice	g_{15}	3	4	0.6	0,58	1	4			0.2		0.1
Scientific computing (Numerical methods)	g_{16}	0	0			0	5				0.2	
Legal/professional/ethics/society	g_{17}	2	4	0.4	0.66	2	4			0.4		
Information systems development	g_{18}	1	3	0.2	0,48	0	2					

(Continued)

Table 6.5 *(Continued)*

Competence		IT		ψ	ω	CS		$x^{(1)}$	$x^{(2)}$	$x^{(3)}$	$x^{(4)}$	$x^{(5)}$
	min	max			min	max					
Analysis of technical requirements g_{19}	3	5	0.6	0.7	2	4	0.3		0.4		0.1
Engineering foundations for SW g_{20}	0	0			1	2			0.2		
Engineering economics for SW g_{21}	0	1			0	1					
Software Modelling and Analysis g_{22}	1	3	0.2	0.56	2	3			0.4	0.2	0.2
Software design g_{23}	1	2	0.2	0.62	3	5			0.6		
Software verification and validation g_{24}	1	2	0.2	0.44	1	2			0.2		
Software evolution (maintenance) g_{25}	1	2	0.2	0.36	1	1			0.2		
Software process g_{26}	1	1	0.2	0.34	1	2					0.2
Software quality g_{27}	1	2	0.2	0.36	1	2					0.2
Computer systems engineering g_{28}	0	0			1	2	0.2				
Digital logic g_{29}	1	1	0.2	0.44	2	3	0.4				
Distributed systems g_{30}	1	3	0.2	0.56	1	3	0.2				
Security: issues and principles g_{31}	1	3	0.2	0.46	1	4	0.2		0.1		
Security: implementation and management g_{32}	3	5	0.6	0.46	1	3	0.2				0.1
Systems administration g_{33}	3	5	0.6	0.4	1	1	0.2		0.1		
Systems integration g_{34}	4	5	0.8	0.54	1	2	0.2		0.1		
Digital media development g_{35}	3	5	0.6	0.3	0	1					
Technical support g_{36}	5	5	1.0	0.34	0	1					

courses. However, new courses are sometimes so similar to those that have been taught so far that introducing them does not require any new competencies.

On the basis of the institutes' competence sets and relations between competencies it is possible to examine whether the institutes have sufficient background strength to obtain type (3) competencies – competencies that are possessed by neither of the institutes. If the background strength of obtaining a type (3) competence is higher than critical level x, the competence can be obtained with no additional cost. It corresponds to stage 1 of the algorithm shown in Fig. 6.5. An example analysis of background strengths of institute P1 (ICAT) is presented in Table 6.6. All values of relations between competencies presented in Table 6.6 were obtained through an expert's judgment, so the values reflect similarity of two academic courses.

$$\beta_9(P_1) = MAX_{j-1}^n(\alpha_j...r_{ij}) = \alpha_5...r_{9.5} = 0.42 > \gamma_9 = 0.32$$

$$\Rightarrow g_9 \in Sk(P_1),$$

$$\beta_{18}(P_1) = MAX_{j-1}^n(\alpha_j...r_{ij}) = \alpha_{34}...r_{18.34} = 0.24 < \gamma_{18} = 0.48$$

$$\Rightarrow g_{18} \notin Sk(P_1),$$

$$\beta_{35}(P_1) = MAX_{j-1}^n(\alpha_j...r_{ij}) = \alpha_4...r_{35.4} = 0.08 < \gamma_{35} = 0.3$$

$$\Rightarrow g_{35} \notin Sk(P_1),$$

$$\beta_{36}(P_1) = MAX_{j-1}^n(\alpha_j...r_{ij}) = \alpha_6...r_{36.6} = 0.32 < \gamma_{36} = 0.34$$

$$\Rightarrow g_{36} \notin Sk(P_1).$$

It can be seen from the above analysis that an institute's P_1's background strength of obtaining g_9 competence is at a sufficient level to obtain it without additional cost. Thus, its competence set $Sk(P_1)$ can be expanded:

$$Sk^{x^{(1)}}(P_1) = \{g_4; g_5; g_6; g_7; g_8; g_9; g_{14}; g_{19};$$

$$g_{28}; g_{29}; g_{30}; g_{31}; g_{32}; g_{33}; g_{34}\};$$

$$x^{(1)} = \{0:3; 0:4; 0:6; 0:4; 0:4; 0:4; 0:4; 0:1;$$

$$0:3; 0:2; 0:4; 0:2; 0:2; 0:2; 0:2\}$$

Table 6.6 Analysis of background strengths of institute P1

r_{ij}		Platform technologies $g_9^{0.4;0.32}$	Information systems $g_{18}^{0.2;0.48}$	Digital media development $g_{35}^{0.6;0.3}$	Technical support $g_{36}^{1;0.34}$
Algorithms and complexity	$g_3^{0.3}$	0.0	0.1	0.2	0.0
Computer architecture and organisation	$g_4^{0.4}$	0.8	0.0	0.2	0.7
Operating systems principle and design	$g_5^{0.6}$	0.7	0.0	0.1	0.5
Operating systems configuration and use	$g_6^{0.4}$	0.8	0.0	0.0	0.8
Net centric principles and design	$g_7^{0.4}$	0.4	0.1	0.0	0.3
Net centric use and configuration	$g_8^{0.4}$	0.5	0.0	0.0	0.5
Information management(DB) theory	$g_{14}^{0.1}$	0.1	0.8	0.0	0.0
Analysis of technical requirements	$g_{19}^{0.3}$	0.5	0.0	0.1	0.8
Computer systems engineering	$g_{28}^{0.2}$	0.1	0.0	0.2	0.1
Digital logic	$g_{29}^{0.4}$	0.0	0.0	0.0	0.0
Distributed systems	$g_{30}^{0.2}$	0.1	0.1	0.0	0.0
Security: issues and principles	$g_{31}^{0.2}$	0.1	0.1	0.0	0.2
Security: implementation and management	$g_{32}^{0.2}$	0.1	0.3	0.0	0.5
Systems administration	$g_{33}^{0.2}$	0.7	0.1	0.0	0.2
Systems integration	$g_{34}^{0.2}$	0.5	0.4	0.0	0.1

After the expansion of P_1 competence, it is necessary to perform a second analysis for the new set $Sk(P_1)$ containing the new competence g_9. The second analysis shows that after the first expansion neither of type (3) competencies can be obtained by P_1 without additional cost.

Similar analyses performed for all institutes proved that neither of them has the background strength necessary to obtain any of type (3) competencies. Competence sets after the first stage of the expansion process are presented in Table 6.7.

Type (1) competencies: $g_1, g_3, g_4, g_5, g_9, g_{14}, g_{17}, g_{22}, g_{23}, g_{24}, g_{25},$ $g_{26}, g_{27}, g_{29}, g_{30}, g_{31}$

Type (2) competencies: $g_2, g_6, g_7, g_8, g_{11}, g_{15}, g_{19}, g_{32}, g_{33}$

Type (3) competencies: g_{18}, g_{35}, g_{36}

6.7.1 Preliminary Expansion Phase

After having adjusted competence sets of all institutes, the preliminary expansion phase should be performed. In this phase, type (2) competencies are assigned to the strongest institute. In this context, the strongest institute is the institute that has the smallest difference between the background strength of obtaining a competence and the critical level of obtaining this competence. The difference is proportional to the cost of obtaining a competence.

During the preliminary expansion phase no new competencies are obtained by the teaching units. Institutes only strengthen their competencies to the levels defined in the requirements (stage 3 of the algorithm). An increase of competence strengths automatically results in an increase of background strengths of type (3) competencies.

Therefore, it is necessary to examine once again whether type (3) competencies can be obtained in result of reaching critical levels of their background strengths (stage 4 of the algorithm).

The analysis of all competence sets carried out after the preliminary expansion phase has shown that institutes P_1 and P_3 had managed to reach the critical level of competence g_{36} – technical support. The expanded competence sets are presented in Table 6.7.

Table 6.7 Adjusted competence sets for all teaching units

Competence		$x^{(1)}$	$x^{(2)}$	$x^{(3)}$	$x^{(4)}$	$x^{(5)}$	ψ	ω
Programming fundamentals	g_1					0.8	0.4	0.78
Integrative programming	g_2	0.3		0.1		0.2	0.6	0.48
Algorithms and complexity	g_3	0.4			0.4	0.8	0.2	0.56
Computer architecture and organisation	g_4	0.6					0.2	0.56
Operating systems principles and design	g_5	0.4				0.3	0,2	0.52
Operating systems configuration and use	g_6	0.4		0.1			0.6	0.6
Net centric principles and design	g_7	0.4		0.2			0.6	0.54
Net centric use and configuration	g_8	0.4		0.1			0.8	0.56
Platform technologies	g_9	0.4					0.4	0.32
Theory of programming languages	g_{10}					0.6		
Human–computer interaction	g_{11}		0.4	0.2			0.8	0.74
Graphics and visualisation	g_{12}		0.2					
Intelligent systems (AI)	g_{13}				0.4			
Information management (DB) theory	g_{14}	0.1		0.4		0.1	0.2	0.48
Information management (DB) practice	g_{15}			0.2		0.1	0.6	0.58
Scientific computing (Numerical methods)	g_{16}				0.2			
Legal/professional/ethics/society	g_{17}			0.4			0,4	0.66
Information systems development	g_{18}						0.2	0.48

(Continued)

Table 6.7 *(Continued)*

Competence		$x^{(1)}$	$x^{(2)}$	$x^{(3)}$	$x^{(4)}$	$x^{(5)}$	ψ	ω
Analysis of technical requirements	g_{19}	0.3		0.4		0.1	0.6	0.7
Engineering foundations for SW	g_{20}			0.2				
Engineering economics for SW	g_{21}							
Software modelling and analysis	g_{22}			0.4		0.2	0.2	0.56
Software design	g_{23}			0.6			0.2	0.62
Software verification and validation	g_{24}			0.2			0.2	0.44
Software evolution (maintenance)	g_{25}			0.2			0.2	0.36
Software process	g_{26}					0.2	0.2	0.34
Software quality	g_{27}					0.2	0.2	0.36
Comp systems engineering	g_{28}	0.2					0.2	0.44
Digital logic	g_{29}	0.4					0.2	0.56
Distributed systems	g_{30}	0.2					0.2	0.46
Security: issues and principles	g_{31}	0.2		0.1			0.2	0.46
Security: implementation and management	g_{32}	0.2				0.1	0.6	0.4
Systems administration	g_{33}	0.2		0.1			0.6	0.54
Systems integration	g_{34}	0.2		0.1			0.8	0.3
Digital media development	g_{35}						0.6	
Technical support	g_{36}						1.0	0.34

Table 6.8 Type (2) competencies assignment

Competence	$x^{(1)}$	$x^{(2)}$	$x^{(3)}$	$x^{(4)}$	$x^{(5)}$	ψ	Teaching unit	Difference c
g_2			0.1		0.2	0.6	P_5	0.4
g_6	0.4		0.1			0.6	P_1	0.2
g_7	0.4		0.2			0.6	P_1	0.2
g_8	0.4		0.1			0.8	P_1	0.4
g_{11}		0.4	0.2			0.8	P_2	0.4
g_{15}			0.2		0.1	0.6	P_3	0.4
g_{19}	0.3		0.4		0.1	0.6	P_3	0.2
g_{32}	0.2				0.1	0.6	P_1	0.4
g_{33}	0.2		0.1			0.6	P_1	0.4
g_{34}	0.2		0.1			0.8	P_1	0.6
Sum								3.6

Type (1) competencies: $g_1, g_2, g_3, g_4, g_5, g_6, g_7, g_8, g_9, g_{11}, g_{14}$, $g_{15}, g_{17}, g_{18}, g_{19}, g_{22}, g_{23}, g_{24}, g_{25}, g_{26}, g_{27}, g_{29}, g_{30}, g_{31}, g_{32}, g_{33}$, g_{34}, g_{35}, g_{36}
Type (2) competencies: $\emptyset$
Type (3) competencies: g_{18}, g_{35}

6.7.2 Main Expansion Phase

The main phase of the expansion process requires finding the minimal cost of obtaining every type (3) competence for each teaching unit (stage 4 of the algorithm). Type (3) competencies and corresponding courses should be assigned according to the minimal cost criterion. Values of the cost function were found using the method presented by Wang (Wang and Wang, 1998). Results of this analysis with graphical are presented in Table 6.8.

The total cost of introducing the IT programme to the examined faculty consists of the costs of two phases of competence expansion. The cost of the preliminary expansion is shown in Table 6.9, while the cost of the main expansion phase is presented in Table 6.10. Summing up, the total cost of competence expansion necessary to introduce the IT programme amounts to $3.6 + 0.105 + 0.033 = 3.738$.

Table 6.9 Competence levels after preliminary expansion phase

Competence		$x^{(1)}$	$x^{(2)}$	$x^{(3)}$	$x^{(4)}$	$x^{(5)}$	ψ	ω
Programming fundamentals	g_1					0.8	0.4	0.78
Integrative programming	g_2			0.1		0.6	0.6	0.48
Algorithms and complexity	g_3	0.3			0.4	0.8	0.2	0.56
Computer architecture and organisation	g_4	0.4					0.2	0.56
Operating systems principles and design	g_5	0.6				0.3	0.2	0.52
Operating systems configuration and use	g_6	0.6		0.1			0.6	0.6
Net centric principles and design	g_7	0.6		0.2			0.6	0.54
Net centric use and configuration	g_8	0.8		0.1			0.8	0.56
Platform technologies	g_9	0.4					0.4	0.32
Theory of programming languages	g_{10}							
Human–computer interaction	g_{11}		0.8	0.2		0.6	0.8	0.74
Graphics and visualisation	g_{12}		0.2					
Intelligent systems (AI)	g_{13}				0.4			
Information management (DB) theory	g_{14}	0.1		0.4		0.1	0.2	0.48
Information management(DB) practice	g_{15}			0.6		0.1	0.6	0.58
Scientific computing (numerical methods)	g_{16}				0.2			
Legal/professional/ethics/society	g_{17}			0.4			0.4	0.66
Information systems development	g_{18}						0.2	0.48

(Continued)

Table 6.9 *(Continued)*

Competence		$x^{(1)}$	$x^{(2)}$	$x^{(3)}$	$x^{(4)}$	$x^{(5)}$	ψ	ω
Analysis of technical requirements	g_{19}	0.3		0.6		0.1	0.6	0.7
Engineering foundations for SW	g_{20}			0.2				
Engineering economics for SW	g_{21}							
Software modelling and analysis	g_{22}			0.4		0.2	0.2	0.56
Software design	g_{23}			0.6			0.2	0.62
Software verification and validation	g_{24}			0.2			0.2	0.44
Software evolution (maintenance)	g_{25}			0.2			0.2	0.36
Software process	g_{26}					0.2	0.2	0.34
Software quality	g_{27}					0.2	0.2	0.36
Comp systems engineering	g_{28}	0.2						
Digital logic	g_{29}	0.4					0.2	0.44
Distributed systems	g_{30}	0.2					0.2	0.56
Security: issues and principles	g_{31}	0.2		0.1			0.2	0.46
Security: implementation and management	g_{32}	0.6				0.1	0.6	0.46
Systems administration	g_{33}	0.6		0.1			0.6	0.4
Systems integration	g_{34}	0.8		0.1			0.8	0.54
Digital media development	g_{35}						0.6	0.3
Technical support	g_{36}	1.0					1.0	0.34

Table 6.10 Cost analysis of the main phase of the expansion process

Unit	$g_{18}^{0.2,0.48}$		$g_{35}^{0.6,0.3}$	
	D_{opt}	$c(D)$	D_{opt}	$c(D)$
P_1	$g_{32}^{0.6} \rightarrow g_{32}^{0.872}$ $\searrow$ $g_{18}^{0.2}$ $\nearrow$ $g_{34}^{0.8}$	0.272	$g_{4}^{0.4} \rightarrow g_{4}^{1.0}$ $\searrow$ $g_{35}^{0.6}$ $\nearrow$ $g_{5}^{0.6} \rightarrow g_{5}^{1.0}$	1.0
P_2	*unreachable*	—	$g_{12}^{0.2} \rightarrow g_{12}^{0.33}$ $\searrow$ $g_{35}^{0.6}$ $\nearrow$ $g_{11}^{0.8}$	0.033
P_3	$g_{14}^{0.4} \rightarrow g_{14}^{0.505}$ $\searrow$ $g_{18}^{0.2}$ $\nearrow$ $g_{15}^{0.6}$	0.105	$g_{11}^{0.2} \rightarrow g_{11}^{1.0} \rightarrow g_{35}^{0.3}$	0.8
P_4	*unreachable*	—	$g_{3}^{0.4} \rightarrow g_{3}^{1.0}$ $\searrow$ $g_{35}^{0.6}$ $\nearrow$ $g_{16}^{0.2} \rightarrow g_{16}^{1.0}$	1.4
P_5	$g_{14}^{0.1} \rightarrow g_{14}^{0.533}$ $\searrow$ $g_{18}^{0.2}$ $\nearrow$ $g_{10}^{0.6}$	0.433	*unreachable*	—

6.8 Summary

The increasing importance of economy based on knowledge results in an urgent need for developing methods and algorithms working at the knowledge level. The competence theory gives an opportunity to create such solutions for educational organisation. Competencies in terms of the modern education system play the role of the main instrument of control and management; see the European Framework for Qualification ("Bologna Working Group on Qualifications Frameworks", 2005).

The approaches to estimating the curriculum modification cost, which have been used so far, allow estimating material and financial resources only. The most important issue missing is the knowledge aspect. In this chapter a practical method of the staff's knowledge estimation in educational organisations is proposed. The results can be used by a decision maker or, on the basis of the proposed algorithm, can be implemented in existing decision support systems in universities.

The method proposed in the chapter supports a decision maker in the process of introducing a new curriculum into an educational organisation consisting of several OUs. On the basis of knowledge assessments of OUs the method allows to estimate the cost of the staff's training necessary to meet the requirements of the new curriculum. The training cost estimates are then used as criteria to assign courses defined by the new curriculum to OUs of the educational organisation. All steps of the proposed approach were illustrated in the case study.

Chapter 7

Incentive Model of a Project Learning Process

7.1 Introduction

The ability to create a project team and manage its knowledge is commonly recognised as one the most important qualities of a knowledge and innovative organisation. The management of knowledge and skills, as well as the management of project competencies , has turned out to be an essential factor influencing the performance of a project process. In this chapter, the methods and tools for incentive modelling for project knowledge management is presented. Also the kind of educational system that can be used as a model of the project learning process aimed at the active behaviour of students in the project development of not only knowledge but also acquiring of competencies is examined.

In open distance learning (ODL) conditions, as an incentive model we consider the scenario of a game (interaction, interplay) between the teacher and the students/project team, conducted in a specific education situation and oriented on performing the actions which allow to raise the level of the students' involvement in subject-

Open Distance Learning: Fundamentals, Developments, and Modelling
Oleg Zaikin
Copyright © 2023 Jenny Stanford Publishing Pte. Ltd.
ISBN 978-981-4877-55-8 (Hardcover), 978-1-003-13261-5 (eBook)
www.jennystanford.com

specified task realisation and to extend the repository with complex tasks performed in the project (Zaikin *et al.*, 2016).

The project learning process in every education situation includes didactic, research and education aspects and takes place at the following levels: cognitive, information and computer-based (Gómez-Pérez *et al.*, 2004). At each of these levels the teacher and the students (participants of project team) have their own roles corresponding to different competencies in the project and involvement intensity. At the cognitive level, assumptions are made and tasks are solved. At the information level, the information is exchanged between the participants of the project learning process. The computer-based level is characterised by repository organisation and the ability to use it. The role of the teacher is to develop an ontological model reflecting the project of the education situation, showing the source of information, formulating tasks and presenting methods and examples of their solutions. All ontological models are stored in the repository (Różewski *et al.*, 2011).

In the discussed approach the project tasks are created on the basis of the ontology and differ in their complexity level (Kushtina, 2006). The proposed scenario assumes that the role of the student is to choose a set of tasks relevant to his/her competence and solve it. The final grade depends on the correctness of the solution and the complexity level of the tasks. The project task solved by the student team and highly graded by the teacher is placed in the repository and will serve as an example solution for other students. All materials stored in the repository are copyrighted. This way the students of the team participate in the didactic activity, and we assume that it will raise their self-esteem, which has a positive influence on learning, meaning that it will be a part of the *student's reward function*. At the same time, filling the repository with a wide spectrum of high-quality solved project tasks gives satisfaction to the teacher, for his/her labours, requiring intelligent efforts for preparing the repository, and this will make up the *teacher's reward function* (Small and Venkatesh, 2000).

The teacher's and the student's interaction with the repository can have a research character. We assume that thematically the content of the repository is in concordance with the teacher's scientific research interests, which causes the appearance and extension of the repository with the tasks differing in the complexity level. For helping to solve tasks stored in the repository, the teacher will pay more attention and spend more time with the students. We can assume that a certain part of the students' participation in common research is a challenge and the obtained results are an extra added value (Miller and Brickman, 2004).

The educational aspect is reflected in the *repository development* as a common success of all participants of the project learning process. Making the material copyrighted shows and visualises the contribution and involvement of each participant of the project. Feeling the synergy effect motivates to develop cooperation skills and tolerance. Cooperation in distance conditions requires a more logical formulation of questions and answers. All this reflects the interests of both the teacher and the students (Tuckman, 2007).

7.2 Stating the Incentive Problem in a Specific Education Situation

Game scenario' Teacher- Students' Team' in a Project learning process
The game theoretical model describing the competence-oriented education process can be represented by the following game scenario ("e-Quality", 2004):

1. The *teacher* proposes a set of project tasks for selection, defines the complexity of each task and builds the motivation function for the project team.

2. The *project team* defines the preference function on the set of projects and on the basis of the motivation function and preference function makes a project selection.

3. The *teacher* determines the structure and parameters of the selected project. The project team makes a distribution of the roles relevant to the competence of each participant of the project team. The teacher performs the role of project coordinator.

4. *Each participant of the project team* solves and realises the given complex of project tasks corresponding to each role. The teacher as a member of the project team performs coordination and consultation.

5. The *teacher* after evaluation the executed project tasks assigns rewards for the project, accounting for the income project and the teacher's expenditures of teacher for the project.

6. The *teacher or project team* makes the division of the total reward for the project, accounting for the complexity of executed tasks and individual expenditures of each participant of the project team.

In the proposed game scenario, each participant has the possibility of free choice among the available alternatives. The main result of the game is to extend an ontology graph of the studied subject by means of adding new vertices connected with the existing ones. These new vertices submit new theoretical knowledge and/or skills acquired in performing project tasks. The goal of the teacher is to attract as many project teams to complex project tasks under the following constraints: the total number of hours assigned for the project in scheduling and the time interval allotted for the game (Kushtina, 2006).

Each participant of the project team defines the result of the game in the range of the following features: from the lowest allowable assessment on the project to the highest result for active and successful participation in the project and 'best practices'. The teacher distributes the consultation time among participants of the project team at his/her discretion, but under the condition of the minimum mandatory quota for each student. In this situation, it may be assumed that the teacher is interested to spend the most of his/her time and intellectual resources on joint work with the project. He/she will contribute to update and expand the repository, which provides a basis for compiling the results in the studied subject, with an option to use grants or publications. This aspect motivates the teacher to advance students to more complex work and to spend more resources on their performance. In this case, it is possible to submit the results to a scientific conference. Thus, the proposed scenario allows the teacher and each participant of the project team to formulate the conditions to choose their strategies in the learning process on the basis of individual preferences (Solow *et al.*, 2005).

The teacher's objective function is to maximise the possible extension of the repository through the involvement of a larger number of students for competitive performance of complex project tasks under the restrictions on the total time of consultations.

Considering students' preferences, two extreme cases may be defined to perform project tasks (Różewski and Zaikin, 2015):

a) The students prefer to perform the project tasks in the simplest way, minimising their time and intellectual costs and getting the lowest possible scores. In fact, these students don't participate in the repository extension.

b) The students prefer to participate in the repository extension of knowledge acquired in the project tasks under restrictions to their own time and intellectual resources. Every student intends to use the maximum teacher's time of consultation, competing for it with other participants of the project team.

In case (a) the teacher and students will complete with minimal effort in the learning process, but indicators, such as average achievement and filling of the repository, will be low, although satisfying the low boundary conditions (formal compliance schedule and personal quota for consultation).

In case (b) the teacher and students will spend far more effort, but a great part of successfully completed tasks will be included into the repository and the average achievement will be higher.

At the same time in case (b), learning will be aimed not only at understanding the theoretical material but also at the ability to operate them (i.e. mastering of didactic material at the level of competence).

Thus, the aim of the game is the maximum extension of the repository with original results of the teacher's and students' joint work under restrictions on their time resources and compliance with preferences of all participants of the game (Kusztina *et al.*, 2010).

An incentive problem is considered for a particular learning situation, when we are dealing with special didactic material for particular participants of the learning process (Armstrong, 2015). For this we use the term 'learning situation', which is represented in Fig. 7.1.

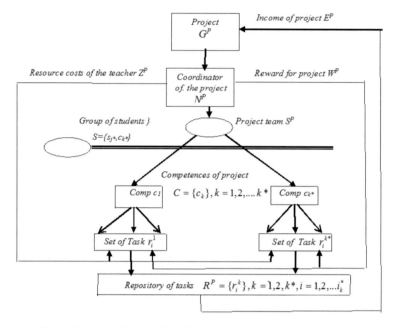

Figure 7.1 Incentive problem for a particular learning situation.

7.3 Formal Model of the Competence-Oriented Project Learning Process

The formal model describing the competence-oriented project learning process has the following four-element structure:

1. *Basic components of the project situation*
   a) Participants of the project learning process $\{G^P, N^P, S^P\}$, where:

   G^P is the domain area of the project
   N^P is the teacher/co-ordinator of the project team
   $S^P = \{s_j\}$ is the group of students/project team
   where $j = 1, 2, \cdots, j^*$ is the index of the student/participant of team
   b) Ontology/hierarchical graph of the project domain area

   $$G^P = \{C^P, L^P\}$$

where:

$C^P = \{c_1^p, c_2^p, \cdots, c_{k^*}^p\}$, $k = 1, \cdots, k^*$ are the competence portions of the project/subordinate nodes of the ontology graph,

$L^P = \{l_1^p, l_2^p, \cdots, l_{k^*}^p\}$ are the hierarchical relations between nodes

c) Set of project tasks $R = \{r_i^k\}$, where

r_i^k is task i consisting of competence portion k

$i = 1, \cdots, i^*$ is the index of the task

$k = 1, \cdots, k^*$ is the index of the acquired competence

d) The level of complexity of a task

$\{q(r_i^k)\}$, $k = 1, \cdots, k^*$, which can be expressed in a binary scale:

$$q(r_i^k) = \begin{cases} 1, & \text{if task } r_i^k \text{ is a complex one,} \\ 0, & \text{otherwise} \end{cases}.$$

e) Time interval assigned to execute the project $t \in [t_0, T_c]$

2. *Decision parameters*

a) Each student/participant of project s_j chooses and executes a certain set of tasks (according to competence k)

$\hat{R}^k(s_j) = \{r_i^k\}$, $i = 1, 2, \cdots, i_k^*$

The project team decision parameter is a binary function of the task set choice $\{r_i^k\}$.

$$y(\{r_i^k\}, s_j) = \begin{cases} 1, & \text{if student } s_j \text{ chooses} \\ & \text{subset of tasks } \{r_i^k\}, \\ 0, & \text{otherwise,} \end{cases}$$

where:

$j = 1, 2, \cdots, j^*$ is the index of the student/participant of the team

$k = 1, 2, \cdots, k^*$ is the index of competencies required for the project

Each student/participant of the project executes a corresponding part/portion of a competence of the project.

There is a one-to-one relation between sets $S = \{s_j\}, j = 1, 2, \cdots, j^*$ and $C = \{c_k\}, k = 1, 2, \cdots, k^*$.

$$S \leftrightarrow C, \ |S| = |C|$$

b) The teacher places up-to-date, complex project tasks in the repository Ω;
$\{q(r_i^k)\}, k = 1, \cdots, k^*, i = 1, \cdots, i^*$ is the level of topicality of a project task, which can be expressed in a binary scale

$$\{q(r_i^k)\} = \begin{cases} 1, & \text{if the task } r_i^{k'} \text{ is chosen by the teacher} \\ & \text{to develop the repository,} \\ 0, & \text{otherwise,} \end{cases}.$$

Ω is a repository of the projects assigned by the teacher for saving of subject knowledge $\Omega = \{p\}$, where p is the executed project.

3. *Criteria parameters of the project/task*
Let's define the following criteria/parameters of the project/task:
   a) *Income of the project*
   The teacher defines the income of project E^P as the total number of executed project tasks assigned for the repository.

$$E^P = \sum_{k=1}^{k^*} \sum_{i=1}^{i_k^*} y(\{r_i^k\}, s_j) q(r_i^k)$$

   b) *Summary resource of the teacher assigned for the project*
   Time of consultation and evaluation of project, access to didactic material, etc., expended on the project

$$Z^P = \sum_{k=1}^{k^*} \sum_{i=1}^{i_k^*} z(\{r_i^k\}),$$

   where $z\{r_i^k\}$ is the teacher's resources expended on the student s_j for a solution of task set $\{r_i^k\}$.
   c) *Reward/win for the project*

The teacher defines the reward/win for the project as a summary number of points gained by the project team for all executed tasks.

$$W^P = \sum_{k=1}^{k^*} \sum_{i=1}^{i_k^*} w\{r_i^k\} y(\{r_i^k\}, s_j),$$

where $w\{r_i^k\}$ is a win gained by student s_j for a solution of task set $\{r_i^k\}$.

d) *Reward/win for the project for a student/participant of the project*

Distribution of the summary win of the team between participants of the project according to the Shapley criterion

$$W^P = \{w(s_1), w(s_2), \cdots, w(s_{j^*})\}$$

e) *The labor intensity for a student/participant of the project*

$$X(s_j) = \sum_{i=1}^{i_k^*} x(\{r_i^k\}) y(\{r_i^k\}, s_j),$$

where $x\{r_i^k\}$ is expenditures of student s_j on a solution of task set $\{r_i^k\}$.

4. *Objective functions*
    a) *Objective function of the teacher*

An objective function of teacher Φ^N is a total win of the teacher (income of project E^P) in the interval $[t_0, T_c]$ of acquiring of the project, taking into account a restriction on expenditures of teacher Z^P.

$$\Phi^N = E^P = \sum_{k=1}^{k^*} \sum_{i=1}^{i_k^*} y(\{r_i^k\}, s_j) q(r_i^k) = \max_{\overline{W}(r_j^k)}$$

$$\sum_{i=1}^{i_k^*} \sum_{j=1}^{j^*} z(\{r_i^k\}) y(\{r_i^k\}, s_j) \leq Z^P,$$

where:

E^P is the income of the project

Z^P is the available resource of the teacher

b) *Objective function of the project team*

An objective function of project team $\Phi(S^P)$ is a win of the students' team, representing a summary reward for the project, assigned by teacher W^P taking into account the restrictions on the expenditures of each team student for execution of project $x(s_j^k)$.

$$\Phi(S^P) = W^P = \sum_{k=1}^{k^*} \sum_{i=1}^{i_k^*} w(\{r_i^k\}) y(\{r_i^k\}, s_j) = \max_{y(r_i^k, s_j)}$$

$$\sum_{i=1}^{i_k^*} x(\{r_i^k\}) y(\{r_i^k\}, s_j) \leq x(s_j^k), \ j = 1, 2, \cdots, j^*$$

5. *Constraints*

a) Summary teacher's resources (time-related, technical, didactic, staff) offered to students for solving tasks

$$Z^\Sigma = \sum_{i=1}^{i_k^*} \sum_{j=1}^{j^*} z(\{r_i^k\}) y(\{r_i^k\}, s_j) \leq Z^P,$$

Where:

$z(\{r_i^k\})$ are resources assigned to the student for solving task set $\{r_i^k\}$

$y(\{r_i^k\}, s_j) = \{1, 0\}$ is a binary function of choice the task set $\{r_i^k\}$ by student s_j

Z_Σ are summary resources for the project led by the teacher

b) Calendar interval $\tau \in [0, T_0]$, appointed for project

$$\min_j \underline{\tau}(s_j) \geq 0, \ \max_j \overline{\tau}(s_j) \leq T_0,$$

where $\underline{\tau}(s_j)$, $\overline{\tau}(s_j)$ are appropriate moments to start and end solving the project by the student.

7.4 Incentive Mechanism in a Multiple-Agent System

7.4.1 Stimulation in a Simple Organisational System

An organisational system, consisting of one centre and one agent - this is the simplest model for which the task of stimulation as follows. The centre's objective function is the difference between income and the cost of incentives, which are paid to the agent. For such a model the *compensatory incentive system* is optimal, the essence of which is as follows.

The agent receives reparation equal to costs in the case of the plan and reparation equal to zero in all other cases. Then the optimal plan is defined as a plan that maximises the difference between the income of the centre and the expenses of the agent.

7.4.2 Incentive Model of Multiple Systems

Motivation is focussed on the impact behaviour of employees in the Organisational system based on the use of motives driving their activities. Selection of a particular method of motivation depends on the specifics of the organisational system, which is determined primarily by the kind of functional connections between employees and managing staff.

A variety of motives determining the behaviour of the organisational system and the need to ensure its competitiveness on long time intervals leads to the necessity of forecasting the dependence of its final results of selected methods of motivation. This, in turn, leads to the need to develop and use appropriate models by which it would be possible to experiment with a selected incentive mechanism for a particular organisational system. Thus the problem of motivational control for an organisational system is reduced to a solution of an incentive task.

The features of organisational system activities and the content of the motivation model are described in the following terms:

Centre – the governing body that determines the levels of subordination, the substantive content of each employee and the plan of their activity

Plan – quantitative indicators describing the results of the activity of the organisation and its individual employees

Agent – any employee working in accordance with the objective set by the management centre

The structure of the organisational system – the ratio of the 'centre – the number of agents', the presence and type of relationships between agents;

The centre's objective function – a formal description of dependence of a goal, which puts the centre on the activity of employees which it must operate

The agent's objective function – a formal description of the dependence of a result, which an employee gets, on the amount and quality of its action within the activity prescribed by the centre

Costs – material and/or time resources that can be used for activities of the considered organisational structure (total or partial)

Income – quantified results for the agents' activity, such as the number of sales and the number of clients served

Incentive mechanism – a set of rules and actions of centre which it uses to motivate the agents.

Consider the kinds of organisational systems that can be used as a model of an educational process aimed at the active behaviour of students in not only the development knowledge but also the acquiring of competencies.

7.4.3 Incentive Mechanism in Multiple System, Where the Reward of Each Agent Doesn't Depend on the Actions of All Other Agents

Consider the possibility of using this model for more complex systems, when several agents are subordinated to one centre (Shubik, 1991). As is known, a model of any organisational system is described by the following parameters: composition, structure, objective functions, admissible sets and being informed.

Let's define these parameters for the studied system:

Composition – one centre and n agents $s_j \in S, j = \{1, 2, \cdots, n\}$

Structure – hierarchical: top level, the centre; bottom level, all agents

The behaviour of the centre and agents is as follows:

The j-th agent chooses an action $y(s_j)$ from the set $A(s_j)$.

The centre chooses a stimulation of the j-th agent $w(s_j)$, depending on the j-th agent's action.

The centre's (teacher's) objective function is the difference between his/her income E^P and the summary expenses paid to the agents (sharing one's own resources Z^P).

$$\Phi(y, q) = \alpha E^P - \beta Z^P =$$

$$= \alpha \sum_{k=1}^{k^*} \sum_{i=1}^{i_k} y(\{r_i^k\}, s_j) q(r_i^k) - \beta \sum_{k=1}^{k^*} \sum_{i=1}^{i_k} y(\{r_i^k\}, s_j) z(r_i^k)$$

where:

$y = (y_1, y_2, \cdots, y_n) \in A = \Pi_{s \in S} A(s_j)$ is the vector of actions of all agents

$z(r_i^k)$ are resources assigned to the student for solving task (r_i^k)

α, β are weight coefficients

By an objective function of each agent we understand the difference between the reward obtained from the centre $W(s_j)$ and the losses connected to solving task $X(s_j)$.

$$\Phi(s_j) = W(s_j) - X(s_j) = \sum_{i=1}^{i_k} y(\{r_i^k\}, s_j) w\{r_i^k\})$$

$$= \sum_{i=1}^{i_k} y(\{r_i^k\}, s_j) x\{r_i^k\},$$

where:

$w\{r_i^k\}$ is a win gained by student s_j for a solution of task set $\{r_i^k\}$

$x\{r_i^k\}$ is are expenditures of student s_j on the solution of task set $\{r_i^k\}$

At the moment of taking the decision (stimulation function for the centre and choice function for the agent), the goal functions and acceptable actions of all participants are known. The centre has the right of the first move, when it chooses a stimulation function, before the agents, with known stimulation functions, choose activities that optimise their goal functions. The centre's choice of a stimulation function takes place on the basis of a simulation meant to serve in foreseeing random characteristics of the basic students' knowledge and parameters of the process of their arrival. Agents choose their strategies independently and do not exchange information or wins; this signifies that we are dealing with a relational dominant strategy (Gubko and Novikov, 2002).

7.4.4 Incentive Mechanism in a Multi-System, Where the Reward of Each Agent Depends on the Actions of all Other Agents

More complex systems of stimulation are needed when the reward of each agent depends not only on the agent's own actions but also on the actions of other agents (Shapley, 1953).

In this case the centre's objective function $\Phi(w,y)$ is the difference between the values of the functions of income $W^P(y)$/summary reward for the project and cost $Z^P(y)$/summary expenses of a centre for the project.

$$\Phi(w,y) = W^P(y) - Z^P(y)$$

The amount of income received by the centre w and expenses of the centre in the course of its work z depend on the effectiveness of the agents activity y.

The agent's objective function $f_j(w_j,y_j)$ is the difference between the reward gained for the work performed $w_j(y_j)$ and incurred with expenses $x_j(y_j)$.

$$\{f_j(w_j,y_j) = w_j(y_j) - x_j(y_j)\},$$
$$y = (y_1, y_2, \cdots, y_n),$$
$$w = (w_1, w_2, \cdots, w_n)$$

In the course of work, agents have expenses $x = (x_1, x_2, \cdots, x_n)$.
Then the following management task arises for the centre in this situation:

It is necessary to find such a system of incentives which guarantees the maximum value of the objective function at the centre on the set of Nash equilibria of the game agents.

In this statement of task we have the following scenario of the game: The centre makes the first move and tells the agents that the reward of each depends of the vector actions of all agents. Then agents make choices of action. The result of the game is a Nash equilibrium (Nash, 1950).

$$E_N(y) = \left\{ y^N \in A \left| \begin{matrix} w_i(y^N) - x_i(y^N) \geq w_i(y_i, y^N_{-i}) = x_i(y_i, y^N_{-i}) \\ \forall i \in N, \ \forall y_i \in A_i \end{matrix} \right. \right\}$$

Then the objective function centre

$$\Phi(w, y) = \min_{y \in E_N(W, o)} [E(y) - \sum_{i \in N} w_j(y)] \to \max_{w(o)}.$$

As seen from this expression, the behaviour of agents for a given function to stimulate is a Nash equilibrium of the game. In this case, the situation is as follows:

1. The centre sets a plan.
2. Agents perform the centre's plan, and their costs depend not only on their actions but also on the actions of other agents.
3. The centre guarantees payment of actual costs to implement the plan for each agent regardless of what other agents do.
4. The negative restitution (penalty) for agents isn't paid.

In the considered multi-agent system, a set of maxima of the agents' objective function transform to a set of Nash equilibria.

7.4.5 Summary

1. To summarise the aforementioned considerations, competence is a general concept, which defines the ability to perform

different patterns of behaviour the basis of accumulated knowledge and experience. While the qualifications relate to all kinds of formal evidence confirming possessing by a person the specific knowledge and skills, simply put, the qualification is formal evidence of specific competencies.

2. Computer-aided management of the students' team in a project requires the use of a formal model of competence, which enables to quantify the usefulness of the students' team to participate in the project. These definitions and the integrated model of the competence reflect only the nature of the competence and don't provide tools for quantitative analysis of the competence. In cases when exact quantitative analysis of competence is required, it is necessary to rely on a model that will provide mathematical foundations and tools to carry it out. This model can precisely describe the competencies and their comparison, determining the cost of the competence increase and determining the adequacy of the competence of the individual to the aim of the tasks and to solve many other problems of a quantitative nature.

3. The results can be used by a teacher as a decision-maker that consider the possibility of using the incentive model for ODL systems, when several agents/project teams are subordinated to one centre/teacher. On the basis of competence assessments of participants of the project team, the method allows to estimate the cost of the team's reward necessary to meet the cognitive level of the project. The competence expansion costs are then used as criteria to evaluate the project tasks and to assign the reward to participants of the project defined.

Chapter 8

Simulation Experiments as a Ground for Acquiring Competencies in the ODL Environment

A simulation experiment can prove to be a useful ground for providing the necessary skills in a very broad sense: from a problem (task) statement, through problem analysis, mathematical and simulation modelling to conducting a research experiment and interpretation of the results. Simulation modelling is a multi-stage and interactive process. Dividing it into stages is a consequence of qualities the simulation process contains in a research method. Each of the stages requires identification of the type and contents of applicable knowledge, and that fact becomes the foundation for development of a problem-specific task sequence. It is therefore reasonable to claim that precise measuring of the competence acquisition process is equally effective as it is in the traditional learning environment. Using the principles presented herein, it is possible to establish an instance of a laboratory of simulation. The laboratory is based on the use of a typical model repository classified according to Kendall's notation (Kleinrock, 1975). That problem has already been analysed in the context of a procedure for

Open Distance Learning: Fundamentals, Developments, and Modelling
Oleg Zaikin
Copyright © 2023 Jenny Stanford Publishing Pte. Ltd.
ISBN 978-981-4877-55-8 (Hardcover), 978-1-003-13261-5 (eBook)
www.jennystanford.com

personal competence acquisition. The content and order of student-oriented tasks are a result of a widely accepted routine dealing with preparation and formulation of stages in the simulation experiment (Banks *et al.*, 2001; Chung, 2003; Robinson, 2004). Let us examine the particular stages of conducting the simulation procedure in the perspective of the tasks provided to the learning individuals.

8.1 Problem Analysis

At this stage both the limit of the problem domain and exogenous limitations (of time, financial, ecological, ergonomic or other nature) specific to the researched phenomenon, system or object should be identified. Moreover, it is necessary to provide a conclusion if simulation as the research method is applicable. The means for a satisfactory problem statement is a reference model of the phenomenon under investigation. Construction of such a model is key to the entire simulation modelling process, and it directly impacts the process stages, especially the requirements for input data, length of a simulation model development process and its validity and the types of experiments targeting such model. At this stage, the didactic material should contain a collection of reference models among which only a few should qualify for further research using the simulation-based experiment. In such a context, the least sophisticated student-targeted problem is identification of a problem that can be solved using the given mathematical apparatus.

8.2 Problem Statement

A problem statement that serves as a roadmap and task formulation for all of the parties involved in creation of the simulation model prevents a situation of failing to follow the defined research objective. Using the results conducted beforehand, the following are to be determined:

- The research objectives
- The input and output elements of the analysed system

- The analysed system's structure
- Assumptions concerning both the unknown and the uncertain system elements
- Model formulation trade-offs that need to be applied with accordance to the modelled real-time system

One of the most important features of the correctly conducted problem statement process is its advanced abstraction. It implicates that the task should be independent of any simulation environment and technique.

For example, let's review the following problem of creating a corporate network for the purpose of a distributed intangible production system. The following tasks can be formulated:

a) Identification of the corporate network's structure and configuration
b) Optimisation of the workplace's efficiency
c) Identification of the network channel's bandwidth

Among these, only tasks (b) and (c) are likely to require the formulation of a simulation experiment. Task (a), however, is carried out using a mathematical model without resorting to simulation. The student is expected to correctly identify the tasks that will require simulation experiment formulation.

8.3 Formulation of a Mathematical Model

Formulation of a mathematical model for a given process is usually extremely abstract. Its main advantage when compared to the simulation model is computing speed and decreased costs of development. Usually, its main downside is related to its limited computing precision in comparison with the simulation model. More importantly, use of the simulation model is effective in performing verification of the mathematical one. The most straightforward way of performing verification is comparing the results from both models or applying methods of statistical analysis. At this stage it is advisable to use mathematical models based on traditional methods that feature greatly refined classification schemes, such as those found in queuing theory (i.e. Kendall's notation). Kendall's notation is a universal research instrument. In the virtual laboratory it is applied not only to perform classification of mathematical models but also to classify and identify other simulation models. The task aimed at students is carrying out the analysis of the mathematical model and interpretation of its parameters in relation to the observed real-time system.

a) *Defining input data*
There is a very big correlation between the structure of the model and requirements regarding the input data used for the experiment. Together with the change in complexity of the model, the requirement for input data, gathered and processed in a proper way, also changes. For example, if the reference model is defined as a one-phase process, the given parameter – operation duration – is interpreted as one number, whereas when the process is defined as a multi-phase one, it is necessary to define the vector of duration of operation performance in each phase of the process. It is the student's task to track how the changes in the reference model influence the definition of input data. In more complicated tasks it is also necessary to track how the change in the nature and scope of the parameters influences the structure of the mathematical model, which will affect the choice of the verification method.

b) *Algorithmicising and programming of the simulation model*
After properly formulating the research problem and developing the reference model, on the basis of which the task assumptions were built, we can progress to the stage of implementing the simulation model. As was already mentioned, properly defined task assumptions are independent of any simulation environment. This gives freedom in choosing the package in which the building of the simulation model and conducting of simulation experiments will start. The choice can be made between such solutions as GPSS/H, Arena, AutoMod, CSIM, Extend, Micro Saint, ProModel, Deneb/QUEST and WITNESS.

For the needs of the virtual simulation laboratory, in the considered examples the Arena package of Rockwell Software Inc., which is a graphical front-end for the simulation language SIMAN, was used. The main argument for using this package is the fact that SIMAN gives great possibilities for storing and re-using simulation models in a repository developed especially for this purpose. Typical models can be stored in the repository in a parameterised form, due to which their use is brought down to solely downloading a proper model from the repository and defining parameter values. In such a case, the process of conducting a simulation experiment is shortened by the stage of simulation model building. The role of the student at this stage is to skilfully use the SIMAN repository.

c) *Model verification and validation*

Both of the stages constitute the most sophisticated and significant challenges to the model analyst (designer). It should be mentioned that validation should not considered as a merely separate collection of procedures emerging sequentially after each model development phase has been completed, but it should become an integral part of the model development process. Validation performed on the fly during the model development process is considered to be more effective than the one committed once the development has been completed, and is therefore of greater usefulness in detecting model defects. The end result of the validation process is a binary one, and it determines whether the model firmly reflects the analysed real objects. The effect of validation is usually achieved through running an iterative comparison analysis conducted between the simulation model and the actual performance of the real system under investigation. The process is repeated as long as its precision is found acceptable.

The purpose of verification is providing an answer if the model has been developed correctly and if so to what degree. The student's task is to address the following questions: Have all of the experiment assumptions been satisfactorily implemented? Have all of the input parameters and the model's logical been correctly reflected in the analysed model? Can the simulation model be used to verify its mathematical counterpart?

d) *Parameter optimisation*

Optimisation of mathematical model parameters is of a very complex nature and can be executed with the use of various methods of optimisation (Law and Kelton, 2000). The key find here is to identify the optimisation method that is most suitable to the given research problem. Optimisation of the model parameters can be aided with the use of specialised software that contains several optimisation algorithms implemented and a mechanism for their selection. An example of such software are Autostat (the AutoMod suite), OptQuest, SimRunner (the ProModel suite) and Output Analyzer (the Arena software) (Banks *et al.*, 2001). The main goal of the student is to formulate a criterion of optimisation applicable to a given example. The problem of using optimisation software tools is not part of the problem discourse.

e) *Result interpretation*

Interpretation of results is an analysis of output data generated during either the experiment or a series of simulation experiments. Since the greatest number of cases is characterised with the stochastic nature of the output data, their analysis necessitates application of stochastic methods. This stage is concluded with a definite answer addressing the initial research goal identified during the problem statement phase. The form specific to the result interpretation process may be distinctive and relies on both the researched problem's nature and the recipient of the processed results. In addition, it is important to use an adequate concept model. If the research object is a queuing systems network, the result data should be presented in terms characteristic to queuing systems theory and Kendall's notation. The student is asked to present the conclusive results using the terms of the reference model in place.

As shown in the aforementioned description, the simulation experiment can be used as a sort of battleground for individually controlled learning as well as a platform for collective project development. The condition for using the simulation experiment in such a situation is to examine coherence between the terms specific to the given stage of the simulation model development.

8.4 Methodology for Developing a Simulation Model

8.4.1 The Statement of a Simulation Experiment

Development of a simulation model requires an initial problem conceptualisation and specification of the given problem using terms of a relevant domain. Problem conceptualisation may usually be represented by an inconsistent structure and is often implemented using an unsophisticated description formed using natural language. The inherent diversity of concepts in that case and the limited precision of specification of the problem statement create major problems at the stage of modelling. The most complex issue is maintaining accuracy in the process of transformation of concepts of the analysed domain to terms and functions applied in a given simulation tool, that is, to a certain portion of procedural knowledge (Kushtina *et al.*, 2005).

Simulation software usually includes libraries of predefined templates for modelling different problem domains. Unfortunately, in most cases they are fairly limited in scope and applicability and are designed to respond to select most typical problems. On the other hand, they effectively handle entire classes of problems within a scope of a given mathematical topic, that is, fundamental (theoretical) knowledge that is represented by a well-defined structure and concept taxonomy. In such a case it is possible to identify and formulate an official analytical model. Moreover, a mathematical apparatus and formal notation are available for precise analytical task formulation. We can refer to the example of Kendall's notation serving queuing systems theory (Kleinrock, 1975).

Therefore, the stage of a problem statement using the concepts of a universally known mathematical engine and its relevant notation can be seen as less complex in performing transformation of the domain concepts to formulas used in the given simulation software than in the example of direct transformation to the terms of a chosen domain characterised by limited precision of its concepts. That is why the previously mentioned stages of a simulation experiment statement conducted in the environment of a virtual laboratory

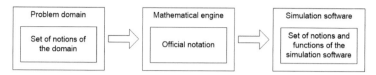

Figure 8.1 Stages of mapping the conceptual model in the process of the simulation experiment statement (Kushtina, 2006).

should be complemented by several iterations of mapping of the verbal problem representation onto the simulation model (Kushtina and Różewski, 2003) (Fig. 8.1). In that context, documents such as the 2010 Mathematics Subject Classification (MSC2010, 2010) in which mathematical taxonomy is introduced play a pivotal role in facilitating the process of identification of an adequate mathematical engine upon formalising the problem domain.

As shown in Fig. 8.1, the simulation experiment statement is exemplified by use of the concepts that are specific to the given simulation software. The concepts form a language for defining the problem and requirements essential for conducting the simulation experiment. The preparation of the concepts is not possible without completing the stage of forming the assumptions about the specific problem using the concepts of the domain that the analysed problem is pertinent to. The simulation experiment statement performed in such an approach is difficult to be translated to the language (terms and formulas) of the simulation software. Hence, an additional modelling process is performed using a mathematical apparatus that is adequate to the given problem. Using that measure significantly simplifies the simulation process of the mathematical model using the given simulation software (Fig. 8.2).

It is vital for the process of formulation of initial assumptions about the researched problem to identify a criterion for performing analysis of the model. Once the criterion has been determined, its examination is required in order to identify its characteristics (continuous or discrete) or to find other significant parameters. The problem containing a properly defined criterion can be subject to modelling using the procedure of mapping the concepts of fundamental knowledge onto procedural knowledge. The mapping can be carried out through identification of direct equivalents between distinct conceptual models (Kushtina *et al.*, 2005).

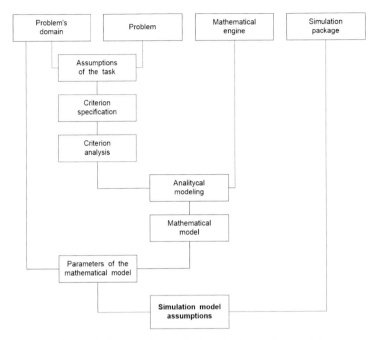

Figure 8.2 Stages of formulation of a simulation model.

8.4.2 The Algorithm for Determining Simulation Experiment Settings

The algorithm for determining the simulation experiment settings will be discussed using an example of a research problem defined using queuing systems theory. Using a problem/statement and the mathematical apparatus of queuing systems theory, an elementary event occurring at the input to the system is to be determined. Naturally, the event is considered the most significant component of the model. In addition, all the parameters involved in the event input stream are to be determined and their average intensity (i.e. a number of events on a given time frame and their characteristic: continuous or discrete). If the input event stream is of stochastic nature, a probability distribution function of time intervals between the incoming events should be determined. In queuing systems theory, a numerous types of events can appear, and for that reason the process of system identification and determination of the operational parameters should be performed for each of the type.

The process of defining the elementary events is of key importance to the queuing systems model. Thus, it should be repeated some number of times in order to thoroughly investigate the given problem.

Once the research activities have been completed, resulting in the definition being sound and acceptable, it then possible to identify the rational laying foundations for the simulation experiment. The assumptions for the experiment are determined using the ARENA simulation software. The previously deliberated mathematical-to-simulation model transformation process is conducted according to the principles of concept mapping (Kushtina *et al.*, 2005).

To conclude, the algorithm for determining a rationale of the simulation experiment for queuing systems theory using the ARENA software is as follows (Fig. 8.3):

1. Given a problem domain, perform a task statement.
2. Determine an elementary event and determine parameters of the event flow.
3. Refine the event definition through recurrent operation with steps (1) and (2). The process is terminated when a researcher is convinced that the problem can be solved using queuing systems theory.
4. Identify a structure of the queuing systems–based system.
5. Determine an initial outline (specification) of the experiment.
6. Create a corresponding simulation model using the ARENA software.

8.4.3 The Process of Adapting the Methodology of the Simulation Experiment to Didactical Purposes

The previously debated stages of the simulation experiment and its composition are commonly applied practices among experts and computer programmers of simulation modelling. To effectively teach those activities carried out within the internet-based environment, a

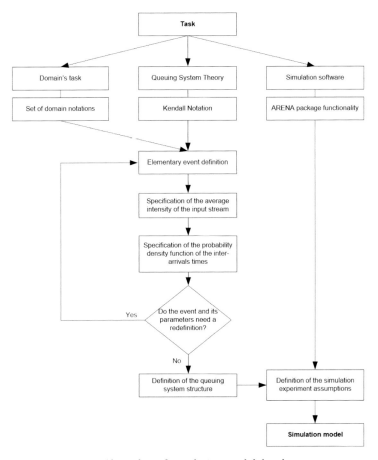

Figure 8.3 Algorithm of simulation model development.

suitable system for teaching computer-aided simulation and an extensive domain knowledge representation method are required. The most significant elements of such system are the repositories storing knowledge related to particular stages of the simulation experiment. The repositories enable multiple and automated knowledge relay to the students enrolled for a course in computer simulation. What's more, a repository offered at each stage of the simulation experiment algorithm is an inherent part of a common knowledge repository in a given teaching system.

Figure 8.4 illustrates a process of defining a simulation experiment. It reflects the procedure that a learning individual upon conducting a simulation experiment needs to maintain while operating in the virtual laboratory.

The following repositories are used in the process of teaching to conduct the simulation experiment (see Fig. 8.4):

- Task repository: stores a variety of tasks and research problems accessed by the student prepared to address and solve them.
- Domain knowledge repository: developed in forms of dictionaries or lexicons containing all the necessary terms required for accurate understanding of a problem by the learning individual. Formalisation of the domain knowledge is carried out on the basis on an extended ontological model.
- Repository of standard queuing systems: contains all of the official queuing system designations classified using Kendall's notation. In addition, it is equipped with concept-mapping patterns.
- Modelling methods repository: stores mathematical modelling methods available to the student.
- Simulation methods repository: in the case of a virtual laboratory of simulation modelling, the repository is strictly related to the ARENA software and stores simulation methods (i.e. software modules with pre-defined parameters) that are provided by the software.
- Analytical models repository: stores analytic methods offered by queuing systems theory that are used to arrange and facilitate an analysis of a queuing system.
- Optimisation methods repository: stores those methods that are widely used at this stage in the laboratory.

While conducting the simulation experiment, the student goes through the following stages:

- *Downloaded task*: The student downloads a task to solve or is assigned one automatically by the system or by the teacher.

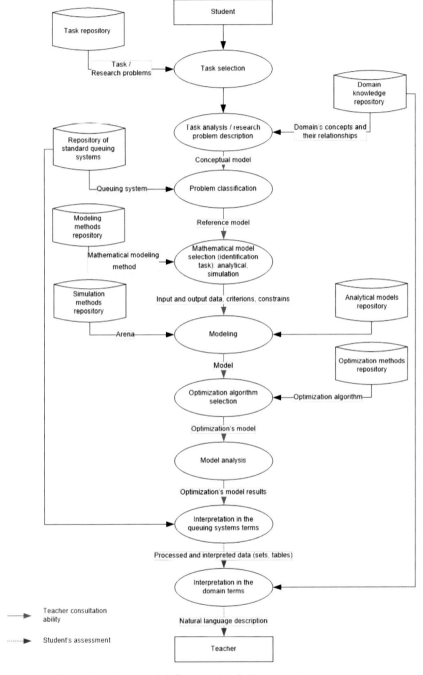

Figure 8.4 Process of defining a simulation experiment.

- *Analysis of the task*: The student performs an analysis of the problem, defines inputs and outputs of the system, defines the structure of the system and identifies the most important parameters. At this stage a domain knowledge repository is useful. It provides definitions of concepts used for formulating the task. The result of this stage is the referential model of the analysed system.

- *Classifying the problem*: If it is possible, the student classifies the analysed system using Kendall's notation.

- *Defining the model type* (*identification*): Using the repository of modelling methods and the self-developed referential model of the studied system, the student chooses a modelling method.

- *Modelling*: Depending on the chosen modelling method, different techniques can be used for building the model (simulation ones, analytical ones or combined ones). Appropriate simulation and analytical methods can be downloaded from the repository.

- *Choosing an optimisation method*: Depending on the type of the developed model and on its characteristics, the student chooses an appropriate optimisation method provided by the optimisation methods repository.

- *Studying the model*: The student conducts one or more experiments on the model in order to obtain results necessary to answer the stated research question.

- *Interpretation of results in the terminology of queuing systems*: Output data obtained as a result of the experiments is interpreted. First of all, the interpretation should be performed in terms of queuing systems, and the data should undergo statistical analysis.

- *Interpretation of results with the use of concepts from the problem domain*: Giving an answer to the task in natural language, using the same concepts that were used to formulate it.

These steps of the procedure are performed in a distance way in the environment of the virtual simulation laboratory; however, after each of them is finished it is possible to consult the teacher who is supervising the learning process.

8.5 Supply Chain in the Learning Management System

8.5.1 Supply Chain and Corporate Network in a Distance Learning Environment

The learning management system (*LMS*) in open distance learning (ODL) is an information system, part of which is an appropriate IT system. The purpose of the IT system is to implement or support the functions of an integrated model of the learning management process. The IT system includes equipment and a dedicated software system. At the stage of conceptualising each IT system, it is necessary to determine the technological level of the equipment and its adaptation to the structure and function of the dedicated software system.

The main conceptual principle of the IT system intended to implement the LMS function is to meet three conditions: openness, functional modularity and the possibility of using in various configurations of the network environment.

- *Openness*, as a consequence of the LMS operating principles, has been considered in the first chapter.
- *Functional modularity*, which results from the hierarchical structure of the LMS, involves the following properties:
  a. Each module belongs to the relevant sub-system.
  b. A module or group of modules implements one of the real processes taking place in the LMS: administrative process, production process or supporting analytical processes.
  c. Modules are parameterised, which means they are orientated to the parameters of the respective processes.

Modularity defined this way allows you to *configure* various IT system implementation options, depending on the set of required functions and hardware capabilities. Each IT system, regardless of its purpose, requires adapting its structure and parameters to processes that are performed in a network environment. Therefore, it is necessary to characterise the functions and processes of the future system using network resources at the conceptualisation stage. The LMS uses a network environment in the form of a corporate network.

A *corporate network* in ODL is defined as a complex of technical, programme and organisational measures intended for managing all types of processes in the LMS, namely administrative, educational and learning processes.

One of the basic function of the LMS is *to provide remote access to the entire nomenclature of educational products and services.* Providing this function is possible only on the basis of the advanced communication infrastructure of the educational organisation and a well-organised corporate network that connects its departments. From this point of view, we will consider an educational institution as an organisation that conducts distance learning and provides educational products and services in accordance with the LMS concept.

Besides the remote access to educational products and learning services, a corporate network must offer the following functions:

1. *Protection and data security*
   The LMS is a system designed for many users, the number of which is very dynamic. This results in high requirements for registration and identification of employees and learners. Data protection and security is performed on three levels: individual user data, departmental data and data of the entire university.

2. *Authorisation of access to knowledge and information resources*
   Information resources are presented in the form of an administrative database, libraries with bibliographic information of websites. Access to each of them is authorised/coded. Each of the registered students has his/her own key by means of which he/she accesses resources. Knowledge resources are implemented in the form of a repository containing didactic materials. Students have access to the repository through an individual curriculum. Specialists involved in the development and their editing have access priority. Securing the access of specialists to the repository implies the preservation of copyrights.

3. *Joint use of distant technical and software resources*

Such resources include the central server, software specialised for scientific research and a virtual laboratory for conducting simulation experiments.

4. *Shared use by many users of expensive devices*
   Such devices include, for example, equipment for preparing releases (digital cameras, digitalises, graphic stations) or printing machines and laser printers.

5. *Organisation of videoconferences and online lectures*, chats and discussion classes and classes using the virtual laboratory
   The corporate network of an educational organisation can be created on the basis of three types of existing computer networks:

   5.1. *A local computer network* at the level of a separate department

   5.2. *A regional metropolitan network* that allows the exchange of production and information resources between the departments of an educational institution or institutions of one region

   5.3. *A global telecommunications network*, connecting several educational organisations on a national or world scale

The theory and practice of developing corporate networks has been devoted to a lot of world literature, but each time there is the problem of formulating the task of optimising the parameters and structure of the network, depending on the conditions of its implementation. The network environment in the LMS is used for the implementation of learning processes, which, depending on the specificity of the educational product, can be divided into two groups. The first group is oriented towards knowledge, which should be mastered by each student during the learning process. The second one is for the preparation of various types of didactic materials, intended for students, teachers (performing a new role), knowledge engineers, field experts and multimedia designers. A common feature of these educational products is the immaterial form of their production.

In the terminology of European projects related to ODL (e.g. ProLearn (PROLEARN, 2004) e-Quality ("e-Quality", 2004)), production processes are:

1) Support for the *student learning sub-process* (LMS), which means organisational, information, technical and curricular support of the *learning/teaching process*

2) Development of learning materials (*learning content management system* (LCMS), which means the design, production and delivery of methodological materials, such as individual curricula, curricular content, methodology for conducting classes and didactic materials: scripts, textbooks, guides and instructions

The supply chain is an environment for the organisation, coordination and implementation of production processes that take place in the LMS. The management object in the corporate network is the information stream, while in the production chain, the management object is the workflow. The characteristic features of workflows are their stochasticity and repeated appearance of similar but not identical events.

The stochasticity of the production process is related to the random time of appearance of orders and a random time of their execution. However, each order of a certain type requires the same technological route or, in other words, the same order of operations carried out at specialised workplaces.

The structure of the supply chain can be presented as an oriented graph. The elements of the supply chain structure are production nodes (vertices of graph) and workflows (edges of graph). There are two basic types of production chain, with an open and a closed structure. Figure 8.5 shows examples of an open supply chain.

In Fig. 8.5, the example the production nodes (vertices of graph) represents the positions of specialists at various stages of preparing didactic materials. The supply chain and corporate network operate in the same environment, but they differ from each other. Typically, a corporate network is considered at the level of the whole enterprise, while the supply chain is considered at the level of individual organisations, where production processes take place, for example educational centres and publishing departments. The basic parameter of the corporate network is the capacity of communication channels, while the basic parameter of the supply chain is the productivity of its production nodes.

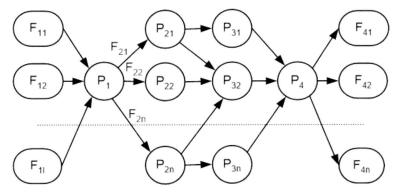

Figure 8.5 Open supply chain in the didactic material preparing system: F, workflows; P, production nodes (source: (Zaikin and Dolgui, 2000)).

8.5.2 An Integrated Model of a Supply Chain in an Educational Organisation

The development of an IT system in the field of education has led to the creation of a new type of educational organisation (a continuing education centre) based on an ODL system. There are several characteristics of an educational organisation, which enable us to recognise it as an production enterprise:

- Dependence of the educational organisation's budget on the paid educational forms and services
- Direct dependence of the educational organisation's competitiveness on the quality of offered programmes and learning outcomes
- Creation of one's own supply chain to organise the learning process and preparation of didactic materials

The characteristic features of the supply chain in an educational organisation are an advanced network infrastructure (on the scale of the region, country, world), a wide range of educational products and services offered, with didactic materials usually in digital form, as well as high-speed equipment and software, used both in the process of preparing didactic materials and in the process of learning students.

These features make it possible to recognise an educational organisation as a new type of production structure, for which the following are characteristic:

a. Stochastic nature of multiply repeated processes, such as arrival of clients' orders, decomposition of orders for separate work, operating of work

b. Unification in one production structure of production nodes with different productivity

c. The ability to handle several works related to different orders at one manufacturing node

All this enables the modelling of this type of production structures as an open or a closed *queuing system* and using of *mass servicing theory* to the analysis and optimisation of the production system parameters.

The processes in an educational organisation can be divided into two types, depending on their designation, source and content of the made works:

1. Processes of design and production of didactic materials
2. Processes of learning management

Learning management processes are related to personal information and therefore require special protection that can be provided by *the corporate network*. The processes of design and production of learning materials are a kind of *intangible production*.

The structure of the production system is distributed due to the following factors:

a. An educational enterprise (EE) may offer a wide range of services to students/clients in the region, the country and even the community of countries.

b. The enterprise has its own distributed structure in which departments, people, equipment and software have been combined.

The parameters of the production system depend on the scale of the extent and the parameters of the processes supported. They are characterised by a quite complicated mutual dependence and high costs to maintain the production system and require their optimal combination. The basis for optimisation is to develop a conceptual model of the production system.

The main goals of developing a conceptual model of a production system are:

a. To define the indices of an EE that have the greatest impact on the structure and parameters of the production system

b. To determine sources and input/output information necessary to develop a production system

c. To select criteria to assess the quality of the production system operation in an EE

d. To outline possible approach to modelling the course of the processes that take place in the production system at the same time

The development of the production system can be made on the basis of various criteria (Buzacott and Shanthikumar, 1993). The following optimisation criteria are most commonly used:

a) Minimising the total time of handling client/student orders counted from the time they are accepted until the completion of the educational product or training service. The sum of operational times of all technological operations and the waiting time in all queues generated in the technological process is assumed as the total time of order processing.

b) Minimising total expenses for technical and software resources of the production system of an EE. These expenses include the costs of their acquisition, installation and operation.

c) Minimising the number of refusals caused by a restriction on the length of the buffer or waiting time when handling customer orders.

As a rule, these criteria are contradictory, and the designer of the production system shapes them, depending on the specific situation.

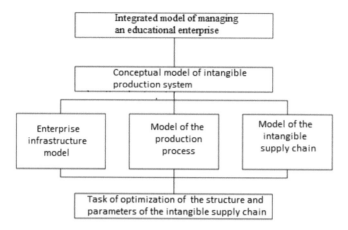

Figure 8.6 Structure of the conceptual model of the intangible production system (source: (Zaikin and Dolgui, 2000)).

8.5.3 The Conceptual Model of the Production System

The conceptual model of the production system is part of an integrated model of managing an EE. The conceptual model of the production system consists of three basic models:

1. The enterprise infrastructure model
2. A model of the production process
3. A model of the intangible supply chain

It is possible to consider these three models within one conceptual model of the production system because they have common variables, such as the rate of the stream of input orders. The discussed models form a conceptual model of the production system, which allows a formal statement of the task of optimising the structure and parameters of the production chain of an EE (Fig. 8.6). We examine each of these components of the conceptual model of the production system.

8.5.3.1 Model of the educational organisation infrastructure

The management problems of an enterprise that implements educational products and services have the following characteristics:

1. Wide nomenclature of educational products and services offered (courses, training, ready knowledge modules, tests, educational games).
2. The branched structure of production processes occurring in a random environment. The following may be random factors: rapidly changing customer demand for educational products and services, the random nature of available educational resources, random time of arrivals of customer requests and random time of operation.

The analysis of random factors shows that the formalisation of the infrastructure of an EE is a problem of large dimensions, which can be interpreted in terms of mass service. The solution of this type of problem using only analytical methods usually leads to serious errors and simplifications – hence the actual problem of combined analytical modelling and simulation methods (Cohen and Lee, 1989).

The objective of simulation modelling is to create a random process of arrival and service of customer orders. In the simulation model, the elementary components of the enterprise's infrastructure are modelled with the preservation of their logical structure and order of creation in real time (Sovetov and Yakovlev, 1998; Shannon, 1978).

Consider the most important elements of this infrastructure in the context of customer service:

1) *Organisational structure of an EE*
   The organisational structure of the enterprise can be presented as a hierarchical graph. The root of the graph is the central office of the enterprise, the intermediate vertices are local departments that support specific areas and the leaves represent end-user terminals. The number of levels in the graph depends on the scale of the enterprise.
2) *Nomenclature of products and services offered by an EE*
   We will determine $\{u_j\} = U$, where:

$j = 1, 1, \cdots, J$ is the product (service) index

U is the complete nomenclature of the enterprise's products and services

3) *Potential area of the EE's activity and its division into territorial units*

We will determine $\{t_k\} = T$, where:

$k = 1, 2, \cdots, K$ is the index of territory covered by the branch O_k

T is the full nomenclature of territories in the area of EE's activity

4) *Rate of inflow of clients' requests*

Let λ_{kj} be the rate of the inflow of customer requests in the territory t_k for products (services); then $\tilde{\tau}_{kj}$ is the average interval between two neighbouring orders for services U_j.

It is obvious that $\lambda_{kj} = \frac{1}{\tilde{\tau}_{kj}}$.

5) *Service channels*

Each department has a certain number of channels providing service and delivery of services to the relevant areas (territories).

6) *Customer population*

The customer population means the number of clients served by the entrprise in a certain area. Due to the population factor, all enterprises can be divided into two types: *finite* and *infinite* populations.

In the first case, the number of clients is comparable to the number of service and delivery servers, while in the second, the number of clients is much higher than the number of service and delivery channels. In the case of an infinite population the rate of arrival of customer requests can be assumed as a constant quantity $\lambda_{kj} = const.$

7) *Distribution of the arrival process of client requests (type of input stream)*

The process of arrival of client requests to territory i for a service j can be described using the distribution function of customer arrival $X_j^i(n, t)$ and rate λ_j^i. This process can be divided into the following classes: deterministic and stochastic, stationary and non-stationary, as well as sequential and batch classes.

8) *Distribution of serving time of customer requests*
The process of customer service during the service j can be described using the distribution function Y_j and average service time $\widetilde{\tau}_j$.

9) *Market segments and competitive enterprises*
The existence of competitive enterprises affects the behaviour of customers when the size of the input queue is limited or the quality of their handling is unsatisfactory.

8.5.3.2 Model of the production process

The model of the production process uses the following basic concepts:

1. *Educational product*
   It is a description of the structure and content of curriculum and didactic materials, taking into account the specificity of distance learning.

2. *Production resources*
   Production resources are used to make an educational product. There are several types of resources needed for the production process:
   - Methodological resources: for learning process realisation
   - Teaching resources: didactic materials
   - Technical resources: equipment and application programmes
   - Information resources: data warehouses and thematic databases
   - Human resources: basic and auxiliary employees (teachers, tutors, operators, administrators)

   Each of these resources is characterised by a fund that represents the volume of the available production resource expressed in natural or temporary units.

3. *Technological operation*

Technological operation is an element of the production process that uses certain production resources. Examples of technological operations in the learning process are registration and identification of students, selection of the subject and curriculum, group formation and testing, teaching a specific subject according to the type of classes and the curriculum, and evaluation (credit, exam, certification). Examples of technological operations in the process of design and production of didactic materials may be publication project, text edition, computer composition and verification.

4. *Production process*

The production process is a chain of operations to acquire knowledge and skills by learners within a specific educational programme using specialised information resources (repository of didactic materials) and production resources (hardware and software, communication channels) and staff (technical, administrative and substantive staff). We distinguish the following production processes: preparation, production and delivery of didactic materials; organisation and management of the learning process; and marketing and development processes, protection of copyright and intellectual property and the process of accreditation and evaluation.

5. *Technological process of learning the subject*

It can be presented as a sequence of operations needed to learn a given subject. Such operations include input knowledge control and conducting classes: lectures, exercises, laboratory work and projects.

6. *State of the production process*

A stochastic vector $\overline{X}(t) = \{x_u(t)\}$ is an average parameter characterising the production process in real time. The component $x_u(t)$ of the vector is the current state of production resources at stage u of the production process, which can be input, intermediate and output.

7. *Time interval*

This is the optimisation interval (functioning) T_0. At time T_0 all production processes are completed (e.g. semester, year).

8.5.3.3 Model of the intangible supply chain

The supply chain in a distributed EE is intended for the production, transmission, delivery and distribution of educational products and services. An intangible supply chain means providing information with delivery channels as a technological operation. By this reason, the supply chain in an education enterprise combines production processes with logistics processes. The same set of parameters characterises both technological and logistic operations. The combination of these operations gives the possibility to consider from one point of view the structure and configuration of the supply chain of an EE.

The production chain is characterised by the following components:

1. *The structure* of the supply chain formally can be presented as an unoriented graph. The vertices of the graph are operating nodes, and the edges are communication channels. Relations between sets of operating nodes and communication channels determine the configuration of the supply chain. Typical configurations of the supply chain are sequential (linear), hierarchical (centralised), network (decentralised) and closed (ring) configurations. Some of these configurations will be discussed in the next sections.

2. *Workflow* in the supply chain is the stream of the works passing in the production process. Workflow management includes determining the technological route/sequence of technological operations, planning, coordinating and synchronising the performance of works in the supply chain. The solution of these tasks must guarantee the optimal value of the criterion of the workflow process management in the supply chain of an EE. The characteristics of the workflow in the supply chain are the distribution of the work arrival process at each operating node and the rate of work arrival.

3. *The operating node* of the supply chain can be presented as a queuing system characterised by the kind of input process, internal structure and output process.

   Figure 8.7 shows the typical structure of an operating node in the production chain.

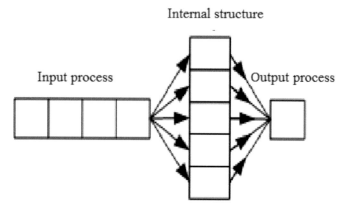

Figure 8.7 Structure of the conceptual model of the intangible production system. The typical structure of the operating node in the supply chain (source: (Zaikin and Dolgui, 2000)).

The operating node is characterised by the following parameters: number of parallel servers, server productivity, expressed in the number of works, or operations performed per time unit and the size of the input buffer, which, together with the number of servers, defines the capacity of the operating node.

4. *Dimensionality of the supply chain* means the total number of works that can simultaneously fit in all nodes of the supply chain. In dependence on the type of buffers, the dimensionality of the supply chain can be infinite or limited (Filipowicz, 1996).

5. Parameters of the quality of operation works in the supply chain. The most important parameters that influence the criterion of the supply chain operation are average time of servicing works in the whole chain, use of equipment and the probability of refusal to perform work, caused by buffer overflows in the operation nodes.

8.6 The Problem of Optimisation of the Structure and Parameters of the Supply Chain of an Educational Organisation

1 *Analysis of the workflow in the intangible supply chain*
There is a big difference between a service and a product in the EE.

The learning service is a customer's order, which concerns the need for learning a certain subject in order to master specific knowledge, skills and abilities. The learning service does not apply to the final product but only to one or several specific jobs, so it does not require a preliminary specification. Usually, this type of order is random and at the same time has specific requirements for specification and quality.

An educational product is a didactic material or a curriculum used at the input of the learning process. The educational product is the result of the process of designing, producing and delivering methodical or didactic material.

The main features of the learning process in the LMS are:

  a. A random time of arrival of the order

  b. A random result of the order decomposition into a set of elementary works

  c. The stochastic nature of the work included in the order

  d. The stochastic nature of time and quality of performance operations

  e. The possibility of running several jobs in parallel

The stochastic nature of the workflow in the supply chain results from the random nature of the arrival time of order on educational services and educational products, the specification of the ordered product (or service) and the volume and quality of the elementary works and operations.

The effective method of analysis this type of processes occurring in the supply chain is based on analytical and simulation modelling, taking into account the stochastic nature of these processes. Effective organisation of the educational process based on modern information technologies requires a new approach. The traditional taxonomy of an educational product (or service) turns out to be insufficient.

In a traditional approach based on such types of educational products as methodologies, scripts, textbooks and training services (lectures, exercises, laboratory, credits, exams), the set of needed resources (employees, hardware, software and order costs) is not clearly defined. In the conditions of implementing new information technologies based on the use of distant human resources, hardware and software, the aforementioned types still play a large role. However, the price and time of order execution strongly depends on many other factors of the educational process, namely individual character of curricula, individual set and order of learning objects in the chosen subject, individual testing and control of knowledge, as well as individual consultations, lectures and laboratories *online*.

The aforementioned educational products and services are characterised by high individuality as well as varying intensity of orders. Each type of product or service is also characterised by a random execution time and a random number of services performed for each item at a specific time interval. The transition from traditional educational products and services to high-tech educational products and services implemented at ODL is similar to the transition from object-oriented production to process-oriented production.

2 *Statement and analysis of the optimisation problems*

In the field of education, we usually deal with complex production processes and a wide range of services and products offered. The problem of the supply chain optimisation for a distributed EE is a complex task of discrete multicriterial optimisation. The area dedicated to the distribution of educational services and products may be extensive (e.g. country, region). The area, in turn, defines the nomenclature of consumers, suppliers, equipment and software. A distributed EE can only be organised if there is a well-developed communication infrastructure (Graves *et al.*, 1998).

In the task of the production chain optimisation, the distribution of consumers of educational services (products) in the territory is treated as known. The territory is understood as the administrative division entity, which is described by the set of parameters, for example geographical location, population density and potential demand for educational products (services). To formalise the problem, it is essential that the population and demands for educational services and products between regions be distributed unequally.

Customer orders for services and educational products can be considered as a real-time process. To formalise the task of optimising the parameters of the supply chain, it is important that the process of arrival of orders be stochastic, stationary and sequential. This means that it can be characterised by its own distribution law. The optimisation task may also contain other random factors, such as the time of servicing of customer demands, the state of hardware and software and the loading of communication channels.

It follows that for a given intensity of customer orders for each type of product or service in each region, it is necessary to deploy production centres (PCs), define each PC configuration and optimise PC productivity by allocating resources.

There are two options to set up an optimisation task:

1) *An open task*, where the entire set of control parameters must be specified (found). This applies to the number of educational centers and the deployment, configuration and performance of each. The open task is difficult to solve, so usually this type of task can be solved only thanks to empirical and common-sense methods.

2) *A closed task*, where a part of the control parameters, for example the number of educational centres, is pre-determined. Only the remaining parameters should be calculated, for example the distribution, configuration and productivity of each centre.

In the closed task, the problem of optimising the supply chain of the distributed EE can be defined in the following form:

For given:
- A set of types of services and educational products offered in the EE
- Distribution and intensity of customer orders arriving for each type of services and products in each region
- Production processes, that is, sequences of operations needed to provide services and products
- Distribution and average time operations

It is necessary to define:

1) Distribution of production centres within the EE's business area

2) Configuration and productivity of each node of the production chain

Criteria:

The minimum value of total production expenditure at the optimisation interval T_0

Decomposition of the optimisation task:

The fundamental problem of optimisation of the supply chain due to its complexity cannot be solved by direct methods. There are two factors that do not allow for its direct solution. The first factor is related to the stochastic nature of processes occurring in the network and other undefined conditions. The second factor is related to the large dimensionality of the task, that is, a great number of parameters that should be included in the optimisation. The dimensionality of the task depends on the set of the control parameters and their combinations. The task of the supply chain optimisation is the NP-full task, because the number of decision variants exponentially depends on the dimensionality of the PC graph.

For this reason, the only way to solve an integral optimisation task is hierarchical, decomposition, that is, the representation of the task in the form of a set of local tasks, solved by direct methods. The stepwise decomposition of the general optimisation task criterion allows defining the optimisation objectives at every step, and the aggregation of control parameters allows to determine the means to achieve these goals.

System analysis of control parameters makes it possible to solve the integral task of PC optimisation in three stages:

1) Stage 1: Deploying the PC in the area of the EE's activity

2) Stage 2: Minimising connection costs by choosing the optimal configuration of the supply chain

3) Stage 3: Optimising the productivity of the operating centres by allocating resources

Each of the aforementioned tasks can be solved on the basis of the EE's own criteria and control parameters (see Table 8.1). Figure 8.8 presents the relationship between components of the conceptual model of the supply chain of a distributed EE.

Table 8.1 Hierarchical decomposition of the integral task of supply chain optimisation (Zaikin, 2002a)

N	Purpose of the task	Optimisation criterion	Control parameters	Scope of optimisation	Model type	Solution method
1	Distribution of production centres in the EE's business area	Maximum total demand for educational services (products)	Distribution of operating centres between the regions of EE	The life cycle of a distributed EE	Model of integer programming	Algorithm for segmentation of the matrix of mutual relations
2	Selection of the optimal configuration of the supply chain	Minimum connection costs between production centres	Configuration and throughput of supply chain channels	The scope of strategic planning for EE's resources	The minimum coverage model for the structure of the supply chain	Branch and bound algorithm in the salesman's task
3	Optimising productivity of operating nodes in each PC	Minimum total production expenses	The fund of resources allocated to each node of the supply chain	The range of EE's tactical planning	Model of an open multi-stream queuing chain	Simulation modelling of mass service systems

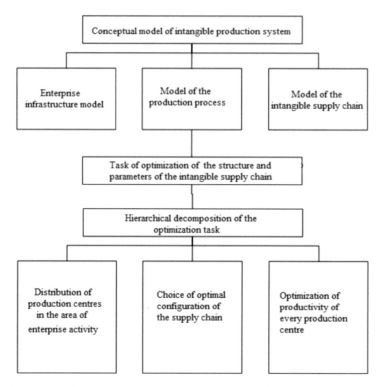

Figure 8.8 Relationships between the components of the conceptual model of the production system and tasks to optimise the EE supply chain (Zaikin, 2002a).

8.7 Queuing Modelling of a Supply Chain in the Learning Management System

8.7.1 Structure of a Closed Supply Chain

Let us consider the process of ODL as a typical closed queuing system containing several special types of servers and one class of learners, hereafter referred to as students (Zaikin and Różewski, 2005). ODL as a queuing system can be described very simply because we can use existing analytical models. For example, we can consider the problem of optimisation of ODL queuing system parameters. To solve this task,

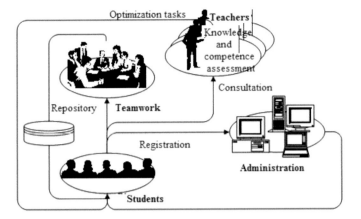

Figure 8.9 Reference model of the ODL closed queuing system (own study).

it is important to learn some of the characteristics typical of queuing systems, such as:

- Average number of students in the class
- Average waiting time in the queue
- Average service time
- Average time thinking for each type of order or the use of servers

Let us consider a situation typical of an EE – a finite population of users, a local network (e.g. intranet) and no restrictions on the length of the waiting queue (Fig. 8.9).

In our deliberations, we will focus on a closed queuing system. The structure of such a system in the context of ODL is presented in Fig. 8.10. In the case of a closed system, we will divide the ODL process into six main components:

1) Server (class) of students
2) Teaching server: a repository of didactic materials
3) Teacher server
4) Simulation server: a repository of simulation models
5) Server of authors

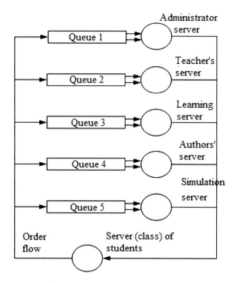

Figure 8.10 Structure of the ODL closed queuing system (own study based on (Zaikin and Dolgui, 2000)).

6) Administrator server: SQL registration and identification database

The student class generates orders with random time intervals. The order process is carried out in the following order:

1) The first student's order is sent to the administrator's server for registration and identification. Each student has the same characteristics of arrival and service.

2) The student generates orders to the teacher's server to solve the input test and receive a theoretical task. The order to the teacher's server can be sent several times.

3) The student, to complete the theoretical task, generates orders for:

   a. Learning server with access to a repository of didactic materials

   b. Teacher's server (for consultation)

   c. Administrator's server (in the case of failure)

4) After the entry test and sending the answer to the theoretical question, the student asks the tutor's server to ask for a practical task, in this case the task with the simulation model.

5) The student generates orders for:

   a. A model repository to conduct a simulation experiment

   b. Tutor's server (for consultation)

   c. Administration server

6) The student leaves the system.

The algorithm for the procedure of passing orders in the learning and control process is shown in Fig. 8.11 and Fig. 8.12.

8.7.2 Statement of the Task

With a large number of students and their independence, the flow of orders generated in the student class and the flow of incoming orders to each type of server can be considered as stationary.

If N is the number of students in the class, then the capacity of the input buffer must not be less than the $N - 1$ for the teacher's server. This is due to the fact that on the one hand, there cannot be more than N orders appearing on the system at the same time, and on the other hand, there cannot be a refusal to serve students.

The process of generating orders of the student class can be described by the Poisson distribution. The service time in the teacher's server is assumed as exponential.

Consider the situation when students need to find didactic materials in the learning server or the relevant exam dates in the administration server. Because in our model the teacher is a consultant, student orders in the learning server are handled in consultation mode.

There are several types of servers and corresponding queuing systems in the ODL environment. Each server has different assignment and productivity. We use Kendall's notation (Kleinrock, 1975) to describe them:

$A/B/X/Y/Z,$

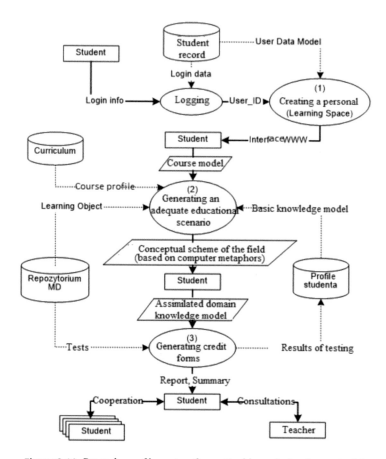

Figure 8.11 Procedure of learning theoretical knowledge (own study).

where:

    A is the distribution of the order arrival process

    B is the service time distribution in the server

    X is the number of parallel servers

    Y is the system capacity

    Z is the queue service discipline

In the queuing system, student works are treated as user orders. Consider a finite population of customers. In general, different students need different times to think. This is related to the cognitive characteristics of each student. In addition, it is important to ensure

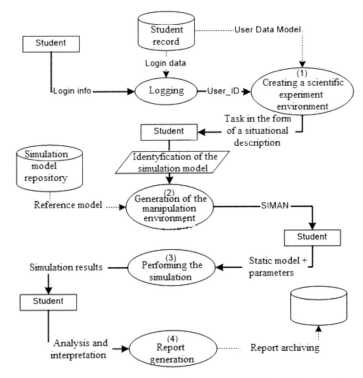

Figure 8.12 Procedure for passing a laboratory task (simulation experiment) (own study).

that the material can be selected and delivered on time. We simplify the real situation and assume that every student has the same cognitive model.

In the case of the ODL system, we distinguish the following queuing systems (Fig.8.13):

1. *Administrator's server M/D/1/N*: Orders from students arrive according to the Poisson process, and the time of service is deterministic (permanent). The number of servers is equal to 1. System capacity is equal to the number of students (N). Servicing discipline is first in, first out (FIFO), that is, first come, first served.

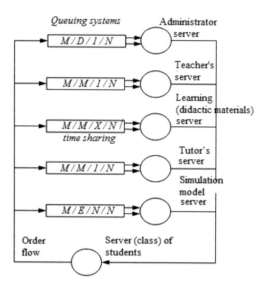

Figure 8.13 Structure of the queuing systems in ODL (own study based on (Zaikin *et al.*, 2000)).

2. *Teacher's server M/M/1/N*: Orders from students arrive in accordance with the Poisson process, and the service time distribution is exponential. The system contains only one teacher. The capacity of the system is equal to the number of students (*N*). Servicing discipline is FIFO.

3. *Learning server (didactic materials) M/M/X/N/ time sharing*: Orders from students arrive in accordance with the Poisson process, and the service time distribution is exponential. The number of servers is equal to *X*, where the size of *X* is determined by economic factors. The capacity of the system is equal to the number of students (*N*). Server operation time is evenly distributed among all orders in the system (time-sharing).

4. *Tutor's server M/M/1/N*: Orders from students arrive according to the Poisson process, and the service time distribution is exponential. The number of servers is equal to 1. The capacity of the system is equal to the number of students (*N*). Servicing discipline is FIFO.

5. *Server of simulation models M/E/N/N*: Orders from students arrive in accordance with the Poisson process, and the service time distribution is Erlang's function. The number of servers is equal to 1. The capacity of the system is equal to the number of students (N). Servicing discipline is FIFO. The Erlang time of service is due to the fact that the process of servicing each task is multi-phase, where k is the number of phases of the simulation experiment. The time of servicing each phase of the experiment can be described by exponential function

6. *Student class M/M/N/N*: Generates orders completely randomly. The time interval between orders is random; both the number of students and the capacity of the system is limited by the number N.

To build a model of the ODL system, it is necessary to determine the average time of staying of orders in queue, average server usage and average time spent in the whole system. In general, the queuing system q_S can be defined by the following parameters: the rate of the arrival of orders λ_S, the intensity of servicing and average time of servicing $\tilde{\tau}_S = 1/\mu_S$, the number of parallel servers N_S, the size of the input buffer b_S and the use of the server ρ_S.

There is a need to specify the number of servers in each queuing system, keeping in mind that each type of server has different productivity and downtime costs. Given next is a formal model for allocating resources in a closed queuing system.

Input data

N is the number of students in the classroom.

$\nu_i, i = 1, 2, \cdots, I$ is the rate of arrival of a student's orders to the server (from one student).

$\tilde{\tau}_i^s, i = 1, 2, \cdots, I$ is the average servicing time for server s_i.

$\mu_i = 1/\tilde{\tau}_i^s$ is the average rate of servicing for the server s_i.

Control parameters: $X_i, i = 1, 2, \cdots, I$ is the number of machines working in parallel for each of the servers.

Criterion function

The criterion function contains three components:

1. Total processing time of all customer orders arriving per unit of time:

$$T_\Sigma = \sum_i N\nu_i \tau_i^S = \sum_i \lambda_i \tilde{\tau}_i^S (X_i),$$

where:

$\lambda_i = \nu_i N$ is the total rate of arrival of client orders to the server s_i

$\tilde{\tau}_i = \tilde{\tau}_i^w + \tilde{}_i^S$ is the average time of order stay in the server S_i

$\tilde{\tau}_i^w$ is the average waiting time of the order in the server queue S_i, $\tilde{\tau}_i^w = F(X_i)$

$\tilde{\tau}_i^S$ is the average time of servicing the order in the server S_i

2. Total server cost (hardware and software)

$$C_\Sigma = \sum_i X_i c_i,$$

where c_i are the costs of one type server S_i.

3. Total costs of work and server downtime (personnel, hardware and software)

$$U_\Sigma = \sum_i [(1 - \frac{\lambda_i}{\mu_i X_i})\delta_i + \frac{\lambda_i}{\mu_i X_i}\gamma_i],$$

where δ_i, γ_i are expenses caused by work or downtime of server S_i per unit of response time.

There are several methods that allow you to find a compromise between the aforementioned components. These include, among others, the selection of one as a criterion function and the adoption of the rest as restrictions, the introduction of weighted factors and Pareto optimisation.

When solving this task, we will consider the first approach, that is, select the first component T_Σ (processing time of all customer orders) as a criterion, and the next two C_Σ (total server costs) and U_Σ (server downtime) are constraints.

Then the task of modelling the workflow can be formulated as follows:

For a given set of server types, $S_i, i = 1, \cdots, I$, included in the ODL system and parameters of each type server $(N, \nu_i, \tilde{\tau}_i^S, \mu_i)$.

Is need to determine: the number of parallel servers of each type $X_i, i = 1, \cdots, I$ guaranteeing a minimum total time needed to complete all orders received:

$$T_\Sigma = \sum_i N\nu_i(\tilde{\tau}_i^w + \tilde{\tau}_i^S) = min, \ (X_i, \ i = 1, \cdots, I)$$

With the restriction: for total costs of educational resources (personal, equipment and software):

$$\alpha C_\Sigma + U_\Sigma = \alpha \sum_i [X_i c_i + (1 - \frac{\lambda_i}{\mu_i X_i})\delta_i + \frac{\lambda_i}{\mu_i X_i}\gamma_i] \leq C_0$$

8.7.3 Method of Solution

The task formulated in this way is the task of integer programming with a non-linear objective function with non-linear constraints. In the general case, with any distribution of orders' flow in the ODL system and any distribution of servicing time, this task cannot be solved only on the basis of analytical methods. It is necessary to perform an experiment on a simulation model.

Suppose the exam in the ODL system described earlier consists of a fixed number of questions for each student in the class. The total time of the exam may be expressed by the following formula:

$$T_{exam} = qN(\tilde{T}_c + \tilde{T}_R) + q \sum_{k=0}^{N_0} \tilde{T}_A(N),$$

where:

q is the number of questions for each student
$(\tilde{T}_c + \tilde{T}_R)$ is the average time of thinking and response
$\tilde{T}_A$ is the average total time spent by the student in the system
N_0 is the the initial number of students in the class

The average total time $\widetilde{T}_A$ can be expressed as the weighted sum of time spent by the student in each type of server:

$$\widetilde{T}_A = \sum_i \nu_i \widetilde{\tau}_i = \sum_i \nu_i (\widetilde{\tau}_i^w + \widetilde{\tau}_i^S)$$

Define p_i as the probability of referring to the server S_i. From the condition of normalisation $\sum_i p_i = 1$ it follows that

$$p_i = \frac{\nu_i}{\sum_i \nu_i}.$$

In the general case, the rate of the arrival of orders in a closed queuing system can be determined by means of a set of linear equations (*traffic equation*):

$$\lambda_i = \sum_{j=1}^{M} \lambda_j p_{ji} = \lambda_c p_i, \ i = 1, \cdots, I, \ \lambda_c = \sum_{i=1}^{M} \lambda_i,$$

where:

M is the number of servers

$P = \|p_{ij}\|$ is the transition matrix

The average number of orders waiting in the queue of server can be determined by using:

$$E[k_i] = \sum_{k=0}^{k_0} k P[k_i = k];$$

$$P[k_i = k] = \sum_{\widetilde{k} \in S, \, k_i = k} P(\widetilde{k}).$$

In the literature (Gordon and Newell, 1967) it was shown that for this type of closed queuing system the following form of the product, shown later, is true:

$$P(\widetilde{k}) = P(\widetilde{0}) \prod_i \left(\frac{\lambda_i}{\mu_i}\right)^{k_i};$$

$$P(\widetilde{0}) = \left\{\sum_{\widetilde{k} \subset S} \prod_{i=1} \left(\frac{\lambda_i}{\mu_i}\right)^{k_i}\right\}^{-1},$$

where:

$P[k_i = k]$ is the probability of waiting for k_i orders in the queue

$P(\widetilde{k})$ is the probability of state $\widetilde{k}$

$P(\widetilde{0})$ is the the probability of a 'zero' state

8.7.4 Conclusion

1. The ODL system is one of the types of intangible production, which is implemented mostly at the information level and is oriented at individual orders of users of an intangible product. User orders create an event flow that appears in the system at random. Servicing orders in intangible production is also a random process. The basic feature of random flow is the rate of the arrival of orders. For intangible production, which serves a large region (on the national, continent or world scale), an uneven distribution of the rate of user orders in different areas of the served region is important.

2. The basic means of implementing intangible production is the supply chain. In market conditions, an educational institution becomes a distributed EE, in which the supply chain is considered as the environment of the educational process. With a defined structure and state of the EE, the problem arises of the optimal use of production resources that make up the supply chain.

3. Because of complexity of the problem of optimisation the use of resources, a conceptual model of a supply chain must be developed at the first stage, which contains the three most important components: infrastructure model of a distributed EE, model of production process and supply chain model. Each

of the mentioned models is characterised by a set of basic concepts and their mutual relations.

4. The task of the production chain optimisation is a complex and multi-dimensional task. For this reason, the only way to solve the integral task of production chain optimisation is its hierarchical decomposition. Analysis of the parameters of the integral task of the optimisation of the supply chain enables finding a solution in three stages:

   Stage 1: Distribution of supply chain nodes in the EE business area

   Stage 2: Minimising the costs of connection between nodes of the supply chain by selecting the optimal chain configuration

   Stage 3: Optimisation of node productivity by allocating production resources between nodes of the production chain

5. The task of optimising the productivity of production nodes of a distributed EE can be defined in the following form. For a given (i) distribution and rate of customer's orders for each type of product, (ii) technological processes of products, (iii) distribution and average time of technological operations *should be determined* the productivity of each node of the supply chain *guaranteeing a minimum value of total production expenses*. The best method of testing such systems is analytical and simulation modelling of mass service processes.

6. The problem of allocating production resources in the system for production of didactic materials in digital form is formulated as the task of optimisation of the parameters of the open queuing system. The arrival time, number and range of orders in such a system are of a random nature. In an education company, many types of products are manufactured, each of which requires a preliminary specification. In addition, the specification of one type of product, belonging to different orders, can vary significantly, for example by changing the required set of work, volume of work and quality requirements.

7. The objective function depends on stochastic and deterministic variables, such as order arrival pattern, service time distribution and number of parallel servers for each node of the queuing

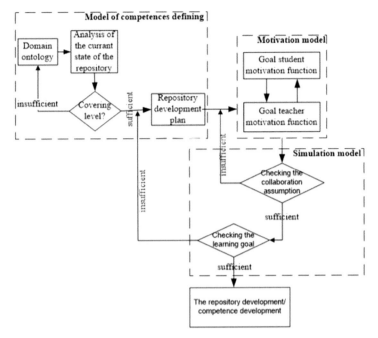

Figure 8.14 Inter-relationship between components (models) of the competence-oriented learning process.

system. The developed algorithm is based on the combined *use of analytical and simulation modelling. Bottleneck* nodes are determined by simulation, and local optimisation of the current bottleneck node is carried out using an analytical model.

8. The management system for ODL processes can be formalised as a closed queuing system, which contains several special types of servers and one class of students. The considered situation is typical for an EE: finite population of users, a local computer network (e.g. intranct) and no restrictions on the length of the waiting queue. The formulated problem is the task of integer programming with a non-linear objective function and non-linear constraints. In general, this task can be solved on the basis of common use of analytical and simulation methods.

8.8 Simulation Model of Teacher and Student Collaboration

The assumed approach to modelling the process of acquiring competencies can be depicted as a set of three components: the model of defining competencies, the motivation model and the simulation model (Malinowska *et al.*, 2013). The inter-relationship between these models is presented in Fig. 8.14.

Integration of the three components allows defining the scope and goal of learning, the conditions of cooperation between the participants of the learning process and the possibilities of analysing the process and influencing the teacher's existing strategy of working with the students. Thus the proposed approach to modelling the process of acquiring competencies appears to be a solution in which:

- The structure of competence and the method of assessing the development of competence (with the use of an ontology graph and a repository) is defined.
- Conditions of competence development are defined (with the use of the motivation model).
- There is a possibility to compare the expected costs that the teacher will bear, considering the assumed repository development (working time) with the results achieved in the didactic process.

The proposed formalisation of cooperation between participants of the learning process during repository complementation indicates that quantitative analysis of this problem from the point of view of expected results and incurred costs is a difficult task. This difficulty is mainly caused by the fact that this process is related to behaviours and preferences of humans, which can change and are difficult to predict. For this reason simulation seems like an adequate mechanism, which allows obtaining information about the analysed object of study and is the basis for further activities related to the functioning of this object.

On the basis of the developed simulation model it is possible to study the adapted *strategy of cooperation between the teacher and a group of students* (Fig. 8.15). The aim of the simulation is to assess

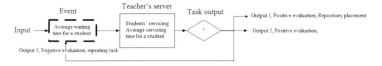

Figure 8.15 General form of the simulation model (own study).

whether the expected repository development plan is possible to realise with the teacher working in a distance mode, with a specified distribution of student arrival and a specified distribution of grading.

Example of a simulation model

Teacher working with a group of students in the process of acquiring competence

An example diagram of a simulation model is presented in Fig. 8.16. It is created in the simulation package Arena made by Rockwell Software (Kelton *et al.*, 1997). To perform simulation experiments, the task of cooperation between the teacher and the students is interpreted as a queuing system, in which students' incoming in ODL is understood as the arrival of tasks in the queuing system and their processing by the teacher as a server with a defined processing time. As a result of the assumed model it will be possible to conduct experiments according to Kendall's notation, for example $M/M/1$, $M/G/1$ (Buzacott and Shanthikumar, 1993; Zaikin, 2002b). In the simulation model it is assumed that students who are being served by the teacher in a given time can acquire a high competence score (source of new resources of the repository), competences an average competence score or be directed for correction. At the same time it is important for all students to be processed in the time planned for acquiring a given competence.

Carrying out simulation experiments can be performed in a simulation package Arena. An example of a simulation model allows considering different values of input parameters and their impact on the learning process.

To determine the value of labour input, information about the level of student motivation can be used, which can be helpful in determining, for example, the time of task execution by a student.

Table 8.2 Examples of conditions for conducting simulation experiments

Simulation Experiment 1	Simulation Experiment 1
Determination of the number of students waiting for service by a teacher at a certain interval of time, at a certain probability distribution outputs, distribution of student services and the estimated time of verification of tasks	Determination of the total time of the teacher assigned for checking of tasks with a certain probability distribution of outputs, distribution of students' service time and the forecasted time of task-checking
Distribution of the arrival of tasks (students) Poisson distribution	Distribution of the arrival of tasks (students) Poisson distribution
Distribution of time for student services: Triangular: minimum – 10 min, most desirable – 15 min, maximum – 30 min	Distribution of time for student services: Triangular: minimum - 10 min, most desirable – 15 min, maximum – 30 min
Number of service channels: 1	Number of service channels: 1
The time interval: 6 days	The time interval: Indefenite
Time assigned for work with students (checking of tasks): 3h/1days	Time assigned for work with students (checking of tasks): 3h/1days
Number of students: 55 students (55 complex tasks)	Number of students: 55 students (55 complex tasks)
Probability of students' going to one of the outputs: completion – 70%, a repository – 15%, correction – 15%;	Probability of students' going to one of the outputs: completion – 70%, a repository – 15%, correction – 15%;
Time to correct the task (delay): 1 day	Time to correct the task (delay): 1 day
Parameters of queue at a server of a teacher	Total time of teacher's job

Table 8.3 Results of simulation experiment

Simulation Experiment 1	
Name of the counter	Achieved value
How much all the tasks checks What do you mean??nauczyciel. POLISH	60
total number of tasks addressed to correction	18
the number of tasks at the entrance.	55
repository	8
output of completed tasks	34
The number of students who at a certain time don't complete the cycle of competence acquisition: **13**	

Simulation Experiment 1	
Name of the counter	Achieved value
how much all the tasks checks a teacher	75
total number of tasks addressed to correction	20
the number of tasks at the entrance.	55
repository	9
output of completed tasks	46
Time interval allowed to complete the cycle of students' competence acquisition: **8 days**	

Figure 8.16 Example diagram of a simulation model of the teacher's work with a group of students in the process of acquiring competence.

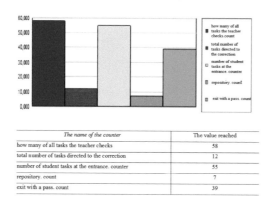

The name of the counter	The value reached
how many of all tasks the teacher checks	58
total number of tasks directed to the correction	12
number of student tasks at the entrance. counter	55
repository. count	7
exit with a pass. count	39

Figure 8.17 Results of the initial simulation experiment.

Simulation experiments can be an indication for the teacher whether the expected strategy of working with the group allows for obtaining the assumed level of competence development. On the basis of the obtained results the teacher can introduce changes to the adopted strategy and check the new conditions in the simulation model. An example of such activity is presented in Table 8.2. For conducting experiments a queuing system model in Kendall's notation $M/G/1$ was used. It was assumed that students (tasks) arrive according to Poisson's distribution (inter-arrival time distribution) and are served by one server (teacher) with triangular service time distribution (Kelton *et al.*, 1997). In Table 8.3 are represented the results of experiment 1 and experiment 2. In Fig. 8.17 and Fig. 8.18 are represented the results of both experiments.

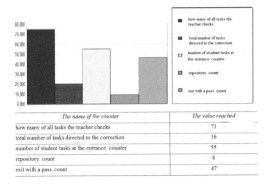

The name of the counter	The value reached
how many of all tasks the teacher checks	71
total number of tasks directed to the correction	16
number of student tasks at the entrance. counter	55
repository. count	8
exit with a pass. count	47

Figure 8.18 Results of a simulation experiment focussed on determining the total time necessary to support all students.

In the case of the first experiment, the adapted strategy of working with the group of students showed that the six-day period assigned by the teacher was insufficient for all students to complete the learning cycle with the specified level of competence (tasks sent to repository, tasks sent to exit with promotion). Looking for a solution to the situation the teacher can for example prolong the time for acquiring the competence, as shown in experiment 2. Such action can simultaneously result in the possibility of filling the repository with additional tasks.

In an illustrative simulation, such parameters are 'The load of a teacher with limited service time on the learning interval', for example, six days, or the expected need to extend this time (if a queue of students who have not completed successfully, the process of acquiring competence will appear).

8.9 Conclusion

1. The model of motivation should be necessary element of the learning management system in conditions of ODL.
2. The model of motivation includes two objective functions (teacher's and students') that define their interest in extension of the knowledge repository.

3. The criterion of efficiency for interaction (cooperation) between teacher and students according to the presented scenario can be the degree of repository extension at a given time.

4. The proposed model of motivation can be solved on the basis of the well-known literature method of Shapley for a cooperative game with a dominant strategy.

Chapter 9

Conclusions

1) Open distance learning

   Distance learning is now a quite widespread technology used in many high schools for better, cheaper and more intensive teaching and learning processes. But what we need in fact now is an open distance learning system (ODLS) because only open resources can be effectively used for achieving civilisation-oriented goals in the whole society. The ODLS as a management object is a complex system that integrates processes of different natures. The way to conduct such system analysis is the theory of hierarchical multi-level systems. As a result were developed the hierarchical structure of the information system for managing an educational enterprise in the conditions of the ODLS and its functional diagram.

   - The hierarchical structure of the information system defines the set of sub-systems and their scope of operation.

   - The functional diagram of the system consists of four nested management cycles and is described as a sequential knowledge processing process. Management on each cycle is carried out using a system decision maker based on specific knowledge and relevant rules.

   - Methods of modelling of competence.

Open Distance Learning: Fundamentals, Developments, and Modelling
Oleg Zaikin
Copyright © 2023 Jenny Stanford Publishing Pte. Ltd.
ISBN 978-981-4877-55-8 (Hardcover), 978-1-003-13261-5 (eBook)
www.jennystanford.com

To summarise the above considerations, competence is a general concept that defines the ability to perform different patterns of behaviour based on accumulated knowledge and experience, while qualifications relate to all kinds of formal evidence confirming possessing by person the specific knowledge and skills. Simply put, qualification is formal evidence of specific competencies. The increasing importance of an economy based on knowledge results in an urgent need for developing methods and algorithms working at the knowledge level. Competence theory gives an opportunity to create such solutions for an educational organisation. Competencies in terms of the modern education system play the role of the main instrument of control and management. The criterion for assessing the quality of the ODL process is the level of competencies acquired by the student over a given period. The development of a mechanism for creating a new specialty profile based on the domain knowledge model and market requirements allows the ODLS to respond timely to labour market requirements. The whole process of adapting programs and the teaching process to changes occurring in the environment of an educational organisation can be interpreted as management oriented on maintaining the chain of information – knowledge – competence. The basis for building the information system to manage the process of acquiring competencies is the repository, which appears as a tool for storing current portions of competencies and monitoring the state of competencies and their increase, and, as a result, organising the process of cooperation between teacher and student during the teaching-learning process.

2) Team project process

The assumed approach to modelling the process of acquiring competencies can be depicted as a set of three components: the model of defining competencies, the motivation model and the simulation model. Integration of the three components allows defining the scope and goal of learning, the conditions of cooperation between the participants of the learning process and the possibilities of analysing the process and influencing the teacher's existing strategy of working with the students.

Thus the proposed approach to modelling the process of acquiring competencies appears to be a solution in which:

- The structure of competence and the method of assessing the development of competence (with the use of an ontology graph and a repository) is defined
- Conditions of competence development are defined with the use of the motivation model
- There is a possibility to compare the expected costs that the teacher will bear considering the assumed repository development (working time) with the results achieved in the didactic process

3) Ontological model

The model presented in the book assumes that defining competencies for the needs of the didactic process needs a mechanism for visualising competencies, which is based on the ontological approach. On the basis of the ontological model, the repository is to be provided and the directions of its further development in cooperation with the teacher and students are defined. An approach focused on the cooperation of participants in the learning process becomes the basis for ontology modelling, under which, on the one hand, goals to be achieved related to the development of competencies in the repository are set and, on the other hand, the possibilities of achieving these goals with time and personal constraints are analysed.

4) Motivation model

The motivation model has to be an obligatory element of an ODLS. It covers the teacher's and the student's motivation functions, which describe their interests in filling the knowledge repository. The measure of success in cooperation between the teacher and students, according to the presented scenario, is the level of repository content expansion in the given time.

Building a repository of students' knowledge can be an important motivational factor for students and for teachers. Students have the opportunity to actively work with their acquired knowledge and competencies in a way where they can follow the results visually and discuss them with their peer-students and teachers. The motivation for the teachers can be an element of the learning management and a pedagogical strategy of ODL. The presented approach developed a new way of teaching and the teacher's involvement; during the process the teacher sees whether the students get the relevant knowledge and competence goals. The students and the teacher can follow the process and relate to the outcome and make the necessary adjustments to the desired goals. Finally it is possible for the teacher to balance the workload with the students' results.

The motivation model includes both the teacher's and the students' interest in extension of the knowledge repository as they both can share and develop the repository into new functions. The measure of success in cooperation between the teacher and the students, according to the presented scenario, is the level of the repository content development in a given time. The proposed motivation model can be solved on the basis of one of the known algorithms realising a cooperative game with dominant strategy.

5) Collaboration model

Computer-aided management of human resources in the project requires the use of formal model of competence, which enables to quantify the usefulness of the research team to participate in the project. These definitions and integrated model of the competence reflect only the nature of the competence and don't provide tools for quantitative analysis of competence. In cases when exact quantitative analysis of competence is required, it is necessary to rely on a model that will provide mathematical foundations and tools to carry it out. This model can precisely describe the competencies, their comparison, determination of the cost of the competence increase, determination of the adequacy of the competence of the individual to the aim of the tasks and the solution of many other problems of a quantitative nature. In the case study a practical method of the team's knowledge estimation in knowledge and innovative organisations is proposed.

The approaches to estimating the curriculum modification cost, which have been used so far, allow estimating material and financial resources only. The most important issue missing is a practical method of the staff's knowledge estimation in educational organisations. The results can be used by a decision maker or on the basis of the organisation consisting of several organisational units. On the basis of knowledge assessments of organisational units, the method allows to estimate the cost of the staff's training necessary to meet the requirements of the new curriculum. The training cost estimates are then used as criteria to assign courses defined by the new curriculum to organisational units of the educational organisation.

6) Incentive model

Computer-aided management of the students' team in the project requires the use of a formal model of competence, which enables to quantify the usefulness of the students' team to participate in the project. These definitions and integrated model of the competence reflect only the nature of the competence and don't provide tools for quantitative analysis of competence. In cases when exact quantitative analysis of competence is required, it is necessary to rely on a model that will provide mathematical foundations and tools to carry it out. This model can precisely describe the competencies, their comparison, determination of cost of the competence increase, determination of the adequacy of the competence of the individual to the aim of the tasks and the solution of many other problems of a quantitative nature.

The results can be used by a teacher as a decision maker, which consider the possibility of using the incentive model for open distance systems, when several agents/project teams are subordinated to one centre/teacher. On the basis of competence assessments of participants of the project team, the method allows to estimate the cost of the team's reward necessary to meet the cognitive level of the project. The competence expansion costs are then used as criteria to evaluate the project tasks and to assign the reward to participants of the project defined.

7) Simulation model

The main idea of simulation modelling is to unify the notions from different knowledge areas, such as the intangible production domain, queuing systems and queuing networks, and simulation experiment. The goal is to create an efficient mechanism for a mathematical-rooted distance learning management process. The motivation model includes both the teacher's and the students' interest in extension of the knowledge repository as they both can share and develop the repository into new functions. This model can be solved on the basis of one of the known algorithms realising a cooperative game. To analyse the constraints of this cooperation, the simulation approach is proposed. The simulation allows to compare the expected costs that the teacher will bear, considering the assumed repository development (working time) with the results achieved in the didactic process (number of students with a high level of competence, participating in the repository, number of students with an average level of competence, etc.). Simulated experiments can be useful indicators for teachers to know whether their pedagogical strategy works for obtaining the assumed level of competence development for the students within a certain amount of work and time.

Glossary

1. Open Distance Learning

Open and Distance Learning (ODL) is a general term for the use of information and communication technologies (ICT) to provide or enhance learning. Around the world, the academic community is discovering and exploring the Internet, ICT teleconferences and related means to achieve an extended learning experience.

Information society is a society where the usage, creation, distribution, manipulation and integration of information is a significant economic, political and cultural activity.

Social aspect of ODL is an innovative movement in education that emerged in the 1970s and evolved into fields of practice and study. The term refers generally to activities that either enhance learning opportunities within formal education systems or broaden learning opportunities beyond formal education systems. Open learning involves classroom teaching methods, approaches to interactive learning, formats in work-related education and training and cultures and ecologies of learning.

Synchronous and asynchronous modes of distance learning are the two most common online learning types. Synchronous distance learning occurs when the teacher and pupils interact in different places but during the same time. Asynchronous distance learning occurs when the teacher and the pupils interact in different places and during different times.

Learning–teaching process comprises the principles and methods used by teachers to enable student learning. These strategies are determined partly on subject matter to be taught and partly by the nature of the learner. For a particular teaching method to be appropriate and efficient, it has to be in relation with the characteristic of the learner and the type of learning it is supposed to bring about.

ODL-based education system is one of the most rapidly growing fields of education nowadays with a substantial impact on all education delivery systems. The new ODL system is growing fast because of the development of Internet-based information technologies, and in particular the World Wide Web. ODL as a social system requires the formulation and appropriate interpretation of certain philosophical values' emergence, synergy, holism and isomorphism

European Credit Transfer System (ECTS) is a standard credit means for comparing the 'volume of learning based on the defined learning outcomes and their associated workload' for higher education across the European Union and other collaborating European countries.

Specialisation – profile – subject may refer to: academic specialisation, may be a course of study or major at an academic institution or may refer to the field in which a specialist practices.

Gantt chart is a type of bar chart that illustrates a project schedule. This chart lists the tasks to be performed on the vertical axis and time intervals on the horizontal axis.

Jackson's network is a class of queueing network where the equilibrium distribution is particularly simple to compute as the network has a product-form solution.

Open distance learning system (ODLS) is divided into three main sub-systems: strategic management system (SMS), learning management system (LMS) and learning content management system (LCMS). Each of the sub-systems includes several modules related to different information system function.

Strategic management system (SMS) provides tools and methods for taking decisions about creating a course basket with consideration of potential students (clients), competitive environment and general social and consumer trends.

Learning management system (LMS) is a software application for the administration, documentation, tracking, reporting and delivery of educational courses, training programs or learning and development programs.

Learning content management system (LCMS) is a software application that can be used to manage the creation and modification of learning content. An LCMS is typically used for enterprise content management (ECM) and web content management (WCM).

Shareable Content Object Reference Model (SCORM) is distance learning standard, the most used international standard in the world for realising distance education courses. SCORM is a set of standards and protocols that make a training content catalogable.

Learning object (LO) is a collection of content items, practice items, and assessment items that are combined on the basis of a single learning objective. The term is credited to Wayne Hodgins and dates from a working group in 1994 bearing the name.

Knowledge repository is a computerised system that systematically captures, organises and categorises an organisation's knowledge. The repository can be searched and data can be quickly retrieved.

Best-practice database is a research, analysis and decision support service for professionals across industries and functions such as marketing, medical affairs, sales, operations, HR, R&D and customer service.

Domain knowledge model (DKM) is a conceptual model of the domain that incorporates both behaviour and data. In ontology engineering, a domain model is a formal representation of a knowledge domain with concepts, roles, data types, individuals and rules, typically grounded in a description logic.

Specialisation profile model (SPM) is formulated by a methodologist (e.g. education officials) within the aims and structures of specialisations oriented on a given domain/market area.

Didactic materials model (DMM) creation of didactic materials is a complex task that needs an effective support for all practitioners involved in such process. This is the goal of the DMM to describe didactic materials requirements like its contents, pedagogical, technical and quality features).

Learning process modelling (LPM) is the activity of representing processes of an educational enterprise, so that the current learning process may be analysed, improved, and automated. LPM is typically performed by information analysts, who provide expertise in the modelling discipline; subject matter experts, who have specialised knowledge of the processes being modelled; or, more commonly, a team comprising both.

Organisation life cycle is the life cycle of an organisation from its creation to its termination. The relevance of a biological life cycle relating to the growth of an organisation, was discovered by organisational researchers many years ago. This was apparent as organisations had a distinct conception, periods of expansion and, eventually, termination

Teaching/learning life cycle is a concept of how people learn from experience. A learning cycle will have a number of stages or phases, the last of which can be followed by the first.

Student's life cycle the sequence of life stages that a student undergoes from incorporation to a university ending with the certification of the competence.

Knowledge-based learning process (KBLP) is a student-centred pedagogy in which students learn about a subject through the experience of solving an open-ended problem found in didactic material. The KBLP does not focus on problem solving with a defined solution, but it allows for the development of other desirable skills and attributes. This includes knowledge acquisition, enhanced group collaboration and communication.

Knowledge acquisition process is a process used to define the rules and ontologies required for a knowledge-based system.

Competence is the ability to use a set of relevant knowledge, skills and abilities to successfully perform 'critical work functions' or tasks in a defined work setting.

Multi-levelled structure of the learning process is a term used to describe the learning process spread vertically between many levels of its management and horizontally across multiple stages of its realisation.

Didactic materials are a conjunction of contents and structural design used to guide learning and teaching processes. Several features such as reusability, semantic intero-perability and collaboration support must be ensured from the earlier stages of the material creation process.

Quality of the learning process is to fulfill learners' needs and expectations. Quality management means coordinated activities to direct and control the education organisation and processes with regard to quality of the learning process.

***The Open University of Catalonia (Universitat oberta de Catalunya - UOC*)** is a private open university based in Barcelona, Spain. The UOC offers graduate and postgraduate programs in Catalan, Spanish and English in fields such as psychology, computer science, sciences of education, information and knowledge society and economics.

2. Methods of Modelling of Competence

Competence is a set of demonstrable characteristics and skills that enable, and improve the efficiency of, performance of a job. Studies on competence indicate that competence covers a very complicated and extensive concept. Competence is also used as a more general description of the requirements of human beings in organisations and communities. If someone is able to do required tasks at the target level of proficiency, they are 'competent' in that area.

Competence-based learning process or competence-based education and training is an approach to teaching and learning more often used in learning concrete skills than theoretical learning. Competence-based learning is a combination of communication, interaction and knowledge production.

Human resource management (HRM) is the strategic approach to the effective management of people in a company or organisation such that they help their business gain a competitive advantage. It is designed to maximise employee performance in service of an employer's strategic objectives.

Personal competence (PC) is self-awareness and self-management, social awareness and relationship management. Before you can begin to plan ways to improve your PC, you need to identify your current level within each competence and then decide the best way you can achieve a 'high' level in each.

Group competence is performing certain activities that are related to job-relevant behaviour, motivation and technical knowledge/skills. The purpose of team collaboration is to assess the effect of team participants' competencies on performance of a learning project.

Qualification is either the process of qualifying for an achievement or a credential attesting to that achievement and may refer to professional social work or professional degrees in social work in various nations.

Competence standards refer to standard terminology used to describe competencies acquired through all stages of the learning process, and the creation of complex procedures for certification of acquired knowledge and skills.

Certification system of competence is a certificate which may be issued to anyone who has successfully completed certain national licenses or has passed an examination to prove the necessary competence.

Project management information system (PMIS) is the coherent organisation of the information required for an organisation to execute projects successfully. A PMIS is typically one or more software applications and a methodical process for collecting and using project information. These electronic systems 'help to plan, execute, and close project management goals'. PMISs differ in scope, design and features, depending upon an organisation's operational requirements.

Project Management Body of Knowledge (PMBOK) – is a set of standard terminology and guidelines (a body of knowledge) for project management. The body of knowledge evolves over time and is presented in the textitGuide to the Project Management Body of Knowledge.

Project management processes are basic project management processes or stages of development used regardless of the methodology or terminology used. Major process groups generally include initiating, planning, implementing, monitoring control and concluding.

Competence-based learning is an approach to teaching and learning more often used in learning concrete skills than abstract learning. Rather than a course or a module, every individual skill or learning outcome (known as a competence) is one single unit. Learners work on one competence at a time, which is likely a small component of a larger learning goal.

Learning profile describes a broad range of human capacities and responsibilities that go beyond academic success. They imply a commitment to help all members of the learning community and learn to respect themselves, others and the world around them.

Labour market refers to the supply and demand for labour in which employees provide the supply and employers the demand.

Required competencies are 'knowledge, skills and abilities required in the labour market', which forms a request for a certain number of workplaces with a certain qualification. Both markets, educational and labour ones, are connected to each other through competencies. Based on these competencies, the labour market forms requirements for the educational market through the required competencies.

Guaranteed competencies are 'knowledge, skills and abilities ensured as a result of the learning/teaching process'. The educational market forms proposals for the labour market through guaranteed competencies.

Competence triplet is a combination of theoretical knowledge, procedural knowledge and project knowledge.

Portion of competence is a connected sub-graph of an ontology graph, formed for each subject (course).

Repository consists of didactical materials for lectures, exercises, laboratory and projects. It imposes new requirements on didactic materials as sources of information, not only on learning topics, but also on teaching and learning strategies, as well as the whole learning process as a combination of communication, interaction and knowledge production.

Repository tasks can be the basis for the development of an e-portfolio of a student and shaping of a personalised career path from the point of view of a student.

3. Team Project Process

Educational philosophy examines the goals, forms, methods and meaning of education. The philosophy of education thus overlaps with the field of education and applied philosophy.

Project process corresponds to the project implementation that is divided into sub-processes. The structure of a project process is defined in the project organisation considering the attainment of corporate objectives and therefore also project objectives. The components of the project process are project start, project preparation, project planning, project implementation and project completion. The two last components should contain also the project decision.

Project team is a team whose members usually belong to different groups, have different functions and are assigned by the project manager to activities for the same project. A team can be divided into sub-teams according to the need.

Students' teamwork is the collaborative effort of a group of students to achieve a common goal or to complete a task in the most effective and efficient way. This concept is seen within the greater framework of a project team, which is a group of inter-dependent individuals who work together towards a common goal.

Acquiring competence of a project team consists of:
- *Intellectual competencies* – understood as a special type of knowledge organisation, considered in terms of conceptual, categorical, semantic abilities and cognitive and personality components of mental activity
- *Practical and project competencies* – what the project team is able to do or accomplish while applying its project management knowledge.

Four basic principles of the education model:
- Motivational environment
- Student activity
- Teamwork
- Structured knowledge

Role of the teacher in the project is vital in education and also the student's life. A person with proper vision, experience and an education degree can enter the teaching profession.

Basic activities of the teacher and university are teaching, research and education.

Traditional learning system, also known as face-to-face, conventional education or customary education, refers to long-established customs that society traditionally used in schools and universities.

Role of didactic materials in ODL is to and collaboration among students, as well as higher thinking skills, such as problem-solving, decision-making and negotiation. Didactic material should satisfy the different learning styles of the students.

Learning situation of ODL is a socially marked situation in which learners are approached as social actors (with a definite status and specific roles) but also live during the period of learning as persons more or less interested or tired, attentive or thinking of something else.

Scenario planning, also called scenario thinking or scenario analysis, is a strategic planning method that some organisations use to make flexible long-term plans.

Motivation function, or incentive/stimulating function, is one of the managerial functions that involves the development and application of various methods of stimulating the participants of a project, including the use of different kinds of incentives.

Motivation model points to the ability of the teacher/project manager to motivate the students in the team towards the common goal of ensuring the success of the project. Motivation models are the approach that the project manager must take to ensure that the team members and the team as a whole are motivated enough to take action and tribute meaningfully to the project.

Team project process includes three stages:

- Identification of a set of competencies required for the project task
- Identification of a set of competencies of the team
- Determining the cost of obtaining the missing competencies required for the project task

Personal competence may appear as a personal property when we are talking about the level of competence of a specialist in a given field. The person needs to identify his/her current level within each competence of the project and then decide the best way to achieve a 'high' level in the project management.

Group/summary competence may appear as a collective trait when we are dealing with a class of problems/tasks to solve, for which we need summary competencies covering several different fields of knowledge.

Human resources management system (HRMS), or human resources information system (HRIS) or human capital management (HCM), is a form of human resources (HR) software that combines a number of systems and processes to ensure the easy management of HR, business processes and data

Expert system is a computer system, in artificial intelligence (AI), that emulates the decision-making ability of a human expert. Expert systems are designed to solve complex problems by reasoning through bodies of knowledge, represented mainly by 'if – then' rules rather than through conventional procedural code.

Strategic management is the identification and representation of the strategies that management implements in order to attain superior results for their organisation, especially in comparison to the competitors in the same industry.

Teaching profile is the subject of teaching, theoretical knowledge and skills related to the use of knowledge. Regardless of what teaching position you are applying for, your career focus, teaching skills and attributes should be customised to the needs of the new teaching profile.

Knowledge management – is the process of creating, sharing, using and managing the knowledge and information of an organisation. It refers to a multi-disciplinary approach to achieving organisational objectives by making the best use of knowledge.

Truly needed competence set is the set of knowledge, information and skills necessary to solve a problem effectively.

Acquired competence set is a set of competencies, i.e. knowledge, information and skills currently possessed

Habitual domain is the space of the problem domain that contains all competencies related to a certain domain area.

Optimal extending process is obtained by using the principle of matching a currently possessed set of competencies to the set of competencies necessary to solve a problem, which minimises the cost at a given stage of extending.

Selecting competent partners is the process of selecting partners to participate in a research project whose purpose is to solve problem E in field D.

Quality management sub-system is a collection of business processes focused on consistently meeting customer requirements and enhancing their satisfaction. It is aligned with an organisation's purpose and strategic direction ISO9001:2015.

European Association for Quality Assurance (ENQA) promotes European cooperation in the field of quality assurance in higher education and disseminates information and expertise among its members and towards stakeholders in order to develop and share good practice and to foster the European dimension of quality assurance.

Rational Unified Process is an iterative software development process framework created by the Rational Software Corporation. RUP is not a single concrete prescriptive process, but rather an adaptable process framework intended to be tailored by the development organisations and software project teams that will select the elements of the process that are appropriate for their needs.

Quality of didactic materials is connected with didactic aspects of educational material creation. There are didactic elements that affect the quality of distance learning.

Mathematics Subject Classification is an alphanumerical classification scheme collaboratively produced by the staff of, and is based on the coverage of, the two major mathematical reviewing databases, Mathematical Reviews and Zentralblatt MATH.

4. Ontology Modelling in Open Distance Learning

Ontology is the philosophical study of being. More broadly, it studies concepts that directly relate to being, in particular becoming, existence, reality, as well as the basic categories of being and their relations.

Ontological model is mathematical groundwork and the corresponding formal model that will be used for the formal representation and validation of methods used in an intellectual modelling system.

'Light' and 'heavy' ontology can be compared to taxonomy and the whole domain, respectively, defining more limitations to domain semantics.

Intensional and extensional definitions are two key ways in which object(s) or concept(s) a term refers to can be defined. The intentional features of a concept represent the content of the concept, i.e. a set of significant features within the domain area, while the extensional ones describes the concept volume, i.e. the number of copies/objects belonging to the class (within the boundaries of the problem, tasks).

Classification of ontologies:
- General (top-level) ontology
- Domain ontology
- Task ontology
- Application ontology

Generalisation of ontology is a form of abstraction whereby common properties of specific instances are formulated as general concepts or claims.

Ontology Engineering Process is a field, in computer science, information science and systems engineering, that studies the methods and methodologies of building ontologies – formal representations of a set of concepts within a domain and the relationships between those concepts.

Languages for description of ontology are formal languages, in computer science and AI, used to construct ontologies. They allow the encoding of knowledge about specific domains and often include reasoning rules that support the processing of that knowledge. Ontology languages are usually declarative languages, are almost always generalisations of frame languages and are commonly based on either first-order logic or on description logic.

Knowledge representation and reasoning is the field of AI dedicated to representing information about the world in a form that a computer system can utilise to solve complex tasks such as diagnosing or dialog in a natural language.

Conceptual knowledge, also referred to as declarative knowledge, refers to the knowledge or understanding of concepts, principles, theories, models, classifications, etc. We learn conceptual knowledge through reading, viewing, listening, experiencing or thoughtful, reflective mental activity.

Procedural knowledge, also known as imperative knowledge, is the knowledge exercised in the performance of some task.

Content and depth of knowledge refers to the body of knowledge and information that teachers teach and that students are expected to learn in a given subject or content area. Depth is defined as the complexity or depth of understanding that is required to answer an assessment question.

Concepts are defined as abstract ideas or general notions that occur in the mind, in speech or in thought. They are understood to be the fundamental building blocks of thoughts and beliefs.

Aggregation (PART_OF), in database management, is a function where the values of multiple rows are grouped together to form a single summary value.

Generalisation (IS_A) is a form of abstraction whereby common properties of specific instances are formulated as general concepts or claims.

Concept taxonomy is the practice and science of classification. The word is also used as a count noun: a taxonomy, or taxonomic scheme, is a particular classification. In a wider, more general sense, it may refer to a classification of things or concepts, as well as to the principles underlying such.

Semantic networks, or frame network, is a knowledge base that represents semantic relations between concepts in a network. This is often used as a form of knowledge representation. It is a directed or undirected graph consisting of vertices, which represent concepts, and edges, which represent semantic relations between concepts mapping or connecting semantic fields.

Learning object is 'a collection of content items, practice items, and assessment items that are combined based on a single learning objective'.

Conceptual network (CN) is a model of natural language under-standing used in AI systems. It investigates how the human brain processes the information, where it stores the results of this process and in which form it represents the results in the outer world.

Conceptual network creation algorithm (CNCA) is intended to identify the knowledge in a specific subject's domain and convert knowledge to the form of a concepts network.

Didactic materials' compilation algorithm (DMCA) adapts the didactic materials for the student by taking into consideration the general educational standards and pedagogical conditions of the learning process.

Semantic relationships are the associations that exist between the meanings of words (semantic relationships at the word level), between the meanings of phrases, or between the meanings of sentences (semantic relationships at the phrase or sentence level).

Case Study 1: Ontological model of RDB of intangible production

Relational database (RDB) is a digital database based on the relational model of data. A software system used to maintain RDBs is a relational database management system (RDBMS). Many RDBMSs have an option of using Structured Query Language (SQL) for querying and maintaining the database.

Ontological model for database management is an ontological approach to managing data using a structure and language consistent with first-order predicate logic.

Information object (entity) reflects a real object, process, event or phenomenon. Qualitative and numerical characteristics of information objects change as a result of real processes. Information objects exist apart from other things, having their own independent existence.

Attribute is a specification (characteristic, distinguishing feature) that defines a property of an information object. It may also refer to a specific value for a given instance. An attribute is characterised by a schema and template. Depending on the nature of the indicated property all attributes in an information object can be divided into: *qualities*, which reflect the qualitative properties of the object, and *bases*, which reflect numerical characteristics of the object.

Complex information unit (CIU) refers to exhaustive characteristics of an information object and can only consist of attributes or other CIUs, Attributes can be associative with one another, forming CIUs.

Key attribute (key) is used to denote the property that uniquely identifies an entity (the Entity Key) and which is mapped to the Primary Key field in a database.

Relation is defined between two information objects at the level of their items and is a binary relation. There are four types of structural relationships: one-to-one, one-to-many, many-to-one and many-to-many.

Functional link is an element of the information search algorithm which shows the order in which the information objects are selected in the processing algorithm.

Structural relationship is a combination of structural relations and function links.

Matrix of relationships, or a matrix diagram or matrix chart, is defined as a management planning tool used for analysing and displaying the relationship between information objects.

Canonical form of the relational model specifies a unique representation for every object, while a normal form simply specifies its form, without the requirement of uniqueness.

Normal form of the relational model specifies a form for every object, without the requirement of uniqueness.

Adjacent matrix, in graph theory, is a square matrix used to represent a finite graph. The elements of the matrix indicate whether pairs of vertices are adjacent in the graph.

Typical queries, in database terms, are used to retrieve data from the database. Queries are one of the things that make databases so powerful. A 'query' refers to the action of retrieving data from your database.

Case Study 2: Ontological Model of the Object-Oriented Database

Personal preference is a learning style or an individual's preferred way of learning. Different people learn in different ways. Each of us has a natural preference for the way in which we prefer to receive, process and impart information.

Bologna Process is a series of ministerial meetings and agreements between European countries to ensure comparability in the standards and quality of higher education qualifications. The process has created the European Higher Education Area under the Lisbon Recognition Convention.

Assessing acquired professional competence is to analyse the topicality of acquired competencies regarding a chosen point of reference. This aspect is especially important when beginning higher education, as it helps in assessing the usability of the knowledge and abilities that are to be acquired in the education process once this process is completed.

Competence comparison model is a set of key competencies necessary for the effective implementation of various duties by employees, all that is necessary to achieve the strategic goals of the company. It describes the knowledge, skills, motives, personal qualities and values of company employees in all positions.

Object-oriented database (OODB) is a database management system in which information is represented in the form of objects as used in object-oriented programming. OODBs are different from RDBs, which are table-oriented. OOBDs have several advantages compared to RDBs: They can store more types of data, access this data faster and allow programmers to reuse objects.

Ontological model of an OODB is based on real-world situations, which are represented as objects, with different attributes. All these objects have multiple relationships between them. The behaviour of the objects is represented using methods. Similar attributes and methods are grouped together using a class.

Strength of competence is a related term of competence. As nouns the difference between strength and competence is that strength is the quality or degree of being strong, while competence is a skill. In psychology, the 'conscious competence' learning model relates to the psychological states involved in the process of progressing from incompetence to competence in a skill.

Case Study 3. Ontological Model of Supply Chain Management

Supply chain (SC) is a system of organisations, people, activities, information and resources involved in moving a product or service from supplier to customer. SC activities involve the transformation of natural resources, raw materials and components into a finished product that is delivered to the end customer.

Extended enterprise is a loosely coupled, self-organising network of firms that combine their economic output to provide products and services offerings to the market. Firms in the extended enterprise may operate independently, for example, through market mechanisms, or cooperatively through agreements and contracts.

Supply chain management (SCM) is the management of the flow of goods and services involves the movement and storage of raw materials, of work-in-process inventory and of finished goods from the point of origin to the point of consumption. Interconnected, inter-related or inter-linked networks, channels and node businesses combine in the provision of products and services required by end customers in an SC.

Production logistics (PL) is generally the detailed organisation and implementation of a production operation. In a general sense, logistics is the management of the flow of products between the point of origin and the point of consumption to meet the requirements of customers or corporations.

Supply chain optimisation is the application of processes and tools to ensure the optimal operation of a manufacturing and distribution supply chain (DSC). This includes the optimal placement of inventory within the SC, minimising operating costs, including manufacturing costs, transportation costs and distribution costs.

Mathematical model of an enterprise is the abstract representation, description and definition of the structure, processes, information and resources of an identifiable business and production processes. A mathematical model of an enterprise consists of the definition of the production process, enterprise environment and distributed production network.

Production process is a complex of phenomena and activities that involve materials and goods gradually undergoing changes.

Production resources (PR) are what is used in the production process to produce output, that is, finished goods and services.

There are four basic resources or factors of production: land, labour, enterprise and capital.

Scheduling is the process of arranging, controlling and optimising work and workloads in a production process or manufacturing process. Scheduling is used to allocate plant and machinery resources, plan HR, plan production processes and purchase materials.

Semantic relationships (entity relationship diagram) describes inter-related things of interest in a specific domain of knowledge. A basic ER model is composed of entity types and specifies relationships that can exist between entities (instances of those entity types).

Management information system (MIS) is an information system used for decision-making and for the coordination, control, analysis and visualisation of information in an organisation.

Distribution supply chain management is an organisational strategy that consists of an integrated approach to planning and managing the entire flow of information about materials, products and services that arise and are transformed in the logistics and production processes of the enterprise.

Ontological model of a DSC is used to test alternative SC decisions, evaluate performance and analyse weaknesses of the SC by an SC analyst or managers. There are three kind of relationships in an ontological model: generalisation, aggregation and structural relation 1:M.

Unified modelling Language (UML) is a general-purpose, developmental, modelling language in the field of software engineering that is intended to provide a standard way to visualise the design of a system.

Mathematical procedures of a DSC include several functions that are planned and managed to improve the total performance of the SC; the general objective is to minimise system-wide costs, while satisfying service level requirements.

Database applications are computer programs whose primary purpose is entering and retrieving information from a computerised database. Early examples of database applications are accounting systems and airline reservation systems. A characteristic of modern database applications is that they facilitate simultaneous updates and queries from multiple users.

5. Motivation Modelling in Open Distance Learning

Learning Motivation is a critical component of learning. Motivation is important in getting students to engage in academic activities. It is also important in determining how much students will learn from the activities they perform. Students who are motivated to learn something use higher cognitive processes in learning about it. Motivation can be a personality characteristic or a stable long-time interest in something.

Didactic material repository considers didactic materials as the conjunction of contents and structural design used to guide learning and teaching processes. Repositories and search engines are those infrastructures that allow the storage, management and retrieval of didactic materials on the basis of their metadata annotation.

Interaction among students and teachers has significantly influenced students' learning experiences, especially in a tutorial setting. A teaching, instead, represents the personal attributes and characteristics that are exhibited by teachers to convey instructions and interact with students.

Activity and cooperation of students and teachers Education is a pedagogical process of interaction. The image of the teacher largely determines the tactics of students in establishing mutual trust in the course of cooperation with subjects of the pedagogical process. In ODL such interaction takes place in conditions of absence of direct contact between the teacher and students.

Cognitive processes use existing knowledge and generate new knowledge. Cognitive processes are analysed from different perspectives within different contexts, notably in the fields of linguistics, neuroscience, psychiatry, psychology, education, philosophy, anthropology, biology, logic, systems and computer science.

Game theory is the study of mathematical models of strategic interaction among rational decision makers. It has applications in all fields of social science, as well as in logic, systems science and computer science. Originally, it addressed zero-sum games, in which each participant's gains or losses are exactly balanced by those of the other participants.

Game situation in ODL is based on the following:

- The teacher's motivation is directly linked to take part in the pedagogical process and interest in sharing their knowledge with the students. Teaches put educational philosophy and objective into the knowledge they transfer to their students.

- The students' motivation is one of the greatest challenges teachers face. Student motivation to learn is an acquired competence developed through general experience but stimulated most directly through modelling, communication of expectations and direct instruction or socialisation.

Motive (the reason of action) is an incentive to act, a reason for doing something, anything that prompts a choice of action, while reason is a cause. As an adjective, motive is causing motion, having the power to move, or tending to move as a motive argument or motive power.

Students' preferences are specific likes and dislikes of an individual human. There are several kinds of personal preferences, each related to different aspects of a student's behaviour Special attention should be paid to each individual's preferences, as they can have a large influence on the decisions that students make and how they behave.

Learning situation is a socially marked situation in which learners are approached as social actors (with a definite status and specific roles), but also live during the period of learning as persons (more or less interested or tired, attentive or thinking of something else).

The ARCS Motivation Model: According to John Keller's ARCS Model of Motivational Design Theories, there are four steps for promoting and sustaining motivation in the learning process: Attention, Relevance, Confidence, Satisfaction.

Dominant strategy, in game theory, is the course of action that results in the highest payoff for a player regardless of what the other player does. A dominant strategy equilibrium is reached when each player chooses his/her own dominant strategy.

Multi-agent system (MAS), or 'self-organised system', is a computerised system composed of multiple interacting intelligent agents. Multi-agent systems can solve problems that are difficult or impossible for an individual agent or a monolithic system to solve.

Virtual laboratory is a project for the development of work for students in physics, chemistry, biology and ecology. Virtual laboratories are implemented using Flash technology. Virtual laboratory products have cognitive value and solve the problem of laboratory work in the absence of the necessary equipment.

Linguistic scales are fuzzy linguistic scales proposed by Cheng, employed to measure the sub-criteria. The fuzzy scalars for these fuzzy linguistic scales are given in Zadeh (1975). Informally, a linguistic variable is a variable whose values are words or sentences in a natural or artificial language.

Linguistic database (LDB) is based on empirical data. An LBD provides a uniform environment for storing data together with its linguistic annotations. It promotes standardised annotation, which facilitates interpretation and comparison of the data. In the motivation model it includes a set of students' motivation parameters, which can be based on the experience of the teacher.

6. Collaboration Modelling in Open Distance Learning

Team collaboration model reveals how people communicate and coordinate to get activities done using communication and IT. If the project focus involves significant collaboration, you would build a collaboration model.

Fuzzy set theory is a research approach that can deal with problems relating to ambiguous, subjective and imprecise judgments, and it can quantify the linguistic facet of available data and preferences for individual or group decision-making.

Collaborative behavioural learning is a situation in which two or more people learn or attempt to learn something together. Unlike individual learning, people engaged in collaborative learning capitalise on one another's resources and skills (asking one another for information, evaluating one another's ideas, monitoring one another's work, etc.). More specifically, collaborative learning is based on the model that knowledge can be created within a team where members actively interact by sharing experiences.

Common habitual domain, for a specific problem/project, which can be treated as a space containing all the competencies related to the project and which can be used in the analysis of the project solution efficiency by different participants of the team with different competencies

Fuzzy subset Z of set Y can be represented by a binary set of values in which the first element corresponds to the value of set Y, and the second one takes values from the interval 0–1. An affiliation function defines membership in a subset of Z. This is a 'degree of affiliation' to a subset of Z (defined values in the interval 0–1).

Power of competence: For each competence c, its power α is a function of the person having it, project P or problem E, in the context of which it is considered.

Base competence : For any pair of competencies a and b in habitual doman (HD) space, if competence b can be achieved with competence a in finite time, then there is relationship between them. In this case, competence a is defined as a base competence for competence b.

Principle of maximum support: If a competence has a lot of base competencies of various powers, then the process of this competence-acquiring is done according to the principle of maximum support.

Organisation theory is a set of propositions that are constructed within the field of organisation science, which includes a study of organisations in practice, and from observation and research develops a body of knowledge that seeks to generalise on the way elements of an organisation interact as well as the way the organisation interacts with its environment.

Games theory is the study of mathematical models of strategic interaction among rational decision-makers. It has applications in all fields of social science, as well as in logic, systems science and computer science. Originally, it addressed zero-sum games, in which each participant's gains or losses are exactly balanced by those of the other participants.

System theory is the inter-disciplinary study of systems. A system is a cohesive conglomeration of inter-related and inter-dependent parts which can be natural or human-made. Every system is bounded by space and time, influenced by its environment, defined by its structure and purpose and expressed through its functioning.

System analysis is 'the process of studying a procedure or business in order to identify its goals and purposes and create systems and procedures that will achieve them in an efficient way (Merriam-Webster dictionary).

Organisational systems are described in many management concepts, theories and practices. From the systems theory point of view every organisation is a complex systems of inter-related parts, and the main task of organisational system is to coordinate the functioning of disparate entities (people, machines, materials, information) to achieve organisational goals.

Theory of hierarchical, multi-level systems is a mathematical theory that investigates the joint behaviour of elements of a multi-level hierarchical structure.

Normative knowledge model is a model in which both human agents and artificial agents have a constructed knowledge about reality, which requires agent active participation, and all knowledge is connected to a knowing agent.

Training cost estimation problem requires defining a certain knowledge representation model that would allow to credibly reflect competencies of personnel in the field of teaching a new course. Another important feature of such a model would be the capability to estimate the cost of competence expansion necessary to reach its required level.

7. Incentive Model Concepts

Incentive problem for a multi-agent active system is solved on the basis of the agents' game decomposition, i.e. by the construction of an incentive system, which implements the optimal (from the principal's point of view) strategies of the agents as the dominant strategies equilibrium (DSE).

Incentive model, in ODL conditions, is a scenario of a game (interaction, interplay) between the teacher and the students/project team, conducted in a specific education situation and oriented on performing the actions which allow to raise the level of the student's involvement in subject-specified task realisation and to extend the repository with complex tasks performed in the project.

Student's reward function is a way to make the students of a team participate in the didactic activity. We assume it will raise his/her self-esteem, which has a positive influence on learning.

Teacher's reward function is filling the repository with a wide spectrum of high-quality solved project tasks, giving satisfaction to the teacher for his/her laborious, intelligent efforts for preparing the repository.

Compensatory incentive system refers to a variety of fields, including education. An incentive system denotes a structure motivating individuals as part of an organisation to act in the interest of the organization. A fundamental requirement of creating a working incentive system for individuals and the organization is understanding human behaviour and motivators of human behaviour.

Relational dominant strategy occurs in game theory when one strategy is better than another strategy for one player, no matter how that player's opponents may play. Many simple games can be solved using dominance. The opposite, intransitivity, occurs in games where one strategy may be better or worse than another strategy for one player, depending on how the player's opponents may play.

Shapley value is a solution concept in cooperative game theory. It was named in honour of Lloyd Shapley, who introduced it in 1953. To each cooperative game it assigns a unique distribution (among the players) of a total surplus generated by the coalition of all players. The Shapley value is characterised by a collection of desirable properties.

Nash equilibria of the game agents, named after the mathematician John Forbes Nash Jr., in game theory, is a proposed solution of a non-cooperative game involving two or more players in which each player is assumed to know the equilibrium strategies of the other players, and no player has anything to gain by changing only his/her own strategy.

8. Simulation Modelling in ODL

Stochastic process, or random process, in probability theory, is a mathematical object usually defined as a family of random variables. Historically, the random variables were associated with or indexed by a set of numbers, usually viewed as points in time, giving the interpretation of a stochastic process representing numerical values of some system randomly changing over time.

Workflow in the supply chain, or flowcharts, show the detailed and specific actions required to achieve end-to-end product delivery. So many actions take place in order for a product to reach the client or consumer – material management, product manufacturing, inventory management, order fulfilment and product delivery.

Simulation modelling is the process of mathematical modelling, performed on a computer, which is designed to predict the behaviour of and/or the outcome of a real-world or physical system. Since they allow to check the reliability of chosen mathematical models, computer simulations have become a useful tool for the mathematical modelling of many natural, social and educational systems.

Simulation experiments are constructed to emulate a physical system. Because these are meant to replicate some aspect of a physical system in detail, they often do not yield an analytic solution. Therefore, methods such as discrete event simulation are used. A computer model is used to make inferences about the system it replicates.

Kendall's notation, in queueing systems theory, a discipline within the mathematical theory of probability, Kendall's notation (or sometimes Kendall notation) is the standard system used to describe and classify a queueing system.

Simulation procedure is used for the design, development, analysis and optimization of technical processes such as educational processes, environmental systems, power stations, complex manufacturing operations, biological processes and similar technical functions.

Reference model, in systems, enterprise and software engineering, is an abstract framework or domain-specific ontology consisting of an inter-linked set of clearly defined concepts produced by an expert or body of experts to encourage clear communication.

Corporate network is a group of computers, connected together in a building or in a particular area, which are all owned by the same company or institutions.

ARENA simulation software is a discrete event simulation and au-tomation software application developed by Systems Modelling and acquired by Rockwell Automation in 2000. It uses the SIMAN processor and simulation language.

Queuing system is used to control queues. Queues of people form in various situations and locations in a queue area. The process of queue formation and propagation is defined as queuing theory.

Closed queuing network consists of several stations. For this structure the flow through each station is passed to other stations. It is a closed network because no flow enters from outside the network.

Traffic equation are equations that describe the mean arrival rate of traffic, allowing the arrival rates at individual nodes to be determined. If the network is stable, the traffic equations are valid and can be solved.

References

Abramowicz, W., Kowalkiewicz, M. and Zawadzki, P. (2002). Tell me what you know or I'll tell you what you know: skill map ontology for information technology courseware, in M. Khosrow-Pour (ed.), *Issues & Trends of Information Technology Management in Contemporary Organizations*, Vol. 1 (IGI Global), pp. 7–10.

ACM and IEEE (2005). Computing Curricula 2005. The Overview Report, https://www.acm.org/binaries/content/assets/education/ curricula-recommendations/cc2005-march06final.pdf

Advanced Distributed Learning Initiative (2001). SCORM 1.2, Sharable Content Object Reference Model, https://www.adlnet.gov/resources/scorm-resources/

Alexander, N. A. (2000). The missing link: an econometric analysis on the impact of curriculum standards on student achievement, *Economics of Education Review* **19**, 4, pp. 351–361.

Allen, J. F. (1983). Maintaining knowledge about temporal intervals, *Communications of the ACM* **26**, 11, pp. 832–843.

Anderson, J. R. (1995). *Learning and Memory: An Integrated Approach* (John Wiley & Sons).

Anderson, J. R. (2000). *Cognitive Psychology and Its Implications* (Worth).

Armstrong, M. (2015). *Armstrong's Handbook of Reward Management Practice: Improving Performance through Reward* (Kogan Page).

Ausubel, D. P. (1968). *The Psychology of Meaningful Verbal Learning: An Introduction to School Learning* (Grune & Stratton).

Ausubel, D. P., Novak, J. D. and Hanesian, H. (1978). *Educational Psychology: A Cognitive View* (Holt, Rinehart and Winston), ISBN 0-03-089951-6, http://lib.ugent.be/catalog/rug01:000005537

Banks, J., Carson II, J. S., Nelson, B. L. and Nicol, D. M. (2001). *Discrete-Event System Simulation*, 5th edn. (Prentice Hall), ISBN 0136062121.

Barker, L. M. (2004). *Psychology* (Pearson Education).

Bates, A. W. (2005). *Technology, e-Learning and Distance Education* (Routledge).

Benjamins, V. R., Fensel, D., Decker, S. and Perez, A. G. (1999). (KA)2: building ontologies for the Internet: a mid-term report, *International Journal of Human-Computer Studies* **51**, 3, pp. 687–712.

Binczewski, A., Meyer, N., Nabrzyski, J., Starzak, S., Stroiński, M. and Węglarz, J. (2001). First experiences with the Polish optical internet, *Computer Networks* **37**, 6, pp. 747–759.

Bologna Working Group on Qualifications Frameworks (2005). The framework of qualifications for the European Higher Education Area, http://www. ehea.info/media.ehea.info/file/WG_Frameworks_ qualification/85/2/ Framework_qualificationsforEHEA-May2005_587852.pdf

Boyatzis, R. E. (1982). *The Competent Manager: A Model for Effective Performance* (John Wiley & Sons).

Brennan, M., Funke, S. and Anderson, C. (2001). The learning content management system: a new elearning market segment emerges, An IDC white paper.

Broens, R. C. J. A. M. and de Vries, M. J. (2003). Classifying technological knowledge for presentation to mechanical engineering designers, *Design Studies* **24**, 5, pp. 457–471, https://doi.org/10.1016/S0142-694X(03)00022-X.

Brophy, J. E. (2010). *Motivating Students to Learn* (Routledge).

Buzacott, J. A. and Shanthikumar, J. G. (1993). *Stochastic Models of Manufacturing Systems*, Vol. 4 (Prentice Hall Englewood Cliffs, NJ).

Buzan, T. and Buzan, B. (1993). *The Mind Map Book* (BBC Books).

Byrne, D. J. and Moore, J. L. (1997). A comparison between the recommendations of computing curriculum 1991 and the views of software development managers in Ireland, *Computers & Education* **28**, 3, pp. 145–154.

Câmara, G., Fonseca, F. and Monteiro, A. M. (2002). Algebraic structures for spatial ontologies, in *II International Conference on Geographical Information Science (GIScience 2002)*, Boulder, CO, AAG.

CANDLE (2003). Collaborative and Network Distributed Learning Environment, 5th Framework Programme: EC, Competitive Project (2000–2003).

Cardy, R. L. and Selvarajan, T. T. (2006). Competencies: Alternative frameworks for competitive advantage, *Business Horizons* **49**, 3, pp. 235–245.

Casati, R. and Varzi, A. C. (1999). *Parts and Places: The Structures of Spatial Representation* (MIT Press).

Chalmers, D. J., In Stich, S. P., and Warfield, T. A. (eds.) (2003). *Blackwell Guide to the Philosophy of Mind, Blackwell.* pp. 102–142.

Chen, T.-Y. (2001). Using competence sets to analyze the consumer decision problem, *European Journal of Operational Research* **128**, 1, pp. 98–118.

Chen, T.-Y. (2002). Expanding competence sets for the consumer decision problem, *European Journal of Operational Research* **138**, 3, pp. 622–648. Chalmers, D. J., In Stich, S. P., and Warfield, T. A. (eds.) (2003). Blackwell Guide to the Philosophy of Mind, Blackwell. pp. 102–142.

Chung, C. A. (2003). *Simulation Modeling Handbook: A Practical Approach* (CRC Press).

Ciszczyk, M. (2006). Problematyka procesu zarządzania kompetencjami, *Metody Informatyki Stosowanej* **10**, pp. 173–179.

Cohen, M. A. and Lee, H. L. (1989). Resource deployment analysis of global manufacturing and distribution networks, *Journal of Manufacturing and Operations Management* **2**, 2, pp. 81–104.

CompTrain (2008). R4.2 *Catalogue of competences - European competence profiles for Multimedia jobs v 1.0.* http://www.annuaire-formation-multimedia.com/catalogue.pdf.

Conole, G., Dyke, M., Oliver, M. and Seale J. (2004). Mapping pedagogy and tools for effective learning design, *Computers & Education*, **43**(1-2), 17–33.

Coombs, C. H., Dawes, R. M. and Tversky, A. (1970). *Mathematical Psychology: An Elementary Introduction* (Prentice-Hall).

Cowan, N. (2001). The magical number 4 in short-term memory: A reconsideration of mental storage capacity, *Behavioral and Brain Sciences* **24**, 1, pp. 87–114.

Crawford, L. (2005). Senior management perceptions of project management competence, *International Journal of Project Management* **23**, 1, pp. 7–16.

De Coi, J. L., Herder, E., Koesling, A., Lofi, C., Olmedilla, D., Papapetrou, O. and Siberski, W. (2007). A model for competence gap analysis, in *Proceedings of 3rd International Conference in WEB Information Systems and technology Barcelona, Spain*, pp. 304–312.

de Sevin, E. and Thalmann, D. (2005). A motivational model of action selection for virtual humans, in *Computer Graphics International 2005*, pp. 213–220.

DeVoe, S. E. and Iyengar, S. S. (2004). Managers' theories of subordinates: a cross-cultural examination of manager perceptions of motivation and appraisal of performance, *Organizational Behavior and Human Decision Processes* **93**, 1, pp. 47–61.

Downes, S. (2001). Learning objects: resources for distance education worldwide, *The International Review of Research in Open and Distributed Learning* **2**, 1.

e-Quality (2004). Quality implementation in open and distance learning in a multicultural European environment, Socrates/Minerva European Union Project (2003–2006), https://www15.uta.fi/projects/e-quality/project.html

Eden, C. (2004). Analyzing cognitive maps to help structure issues or problems, *European Journal of Operational Research* **159**, 3, pp. 673–686.

European Association for Quality Assurance in Higher Education (2016). Guidelines for ENQA agency reviews, https://enqa.eu/indirme/papers-and-reports/occasional-papers/Guidelines%20for%20ENQA%20Agency%20Reviews.pdf

Eynde, F. V. and Gibbon, D. (eds.) (2000). *Lexicon Development for Speech and Language Processing* (Kluwer Academic).

Fandel, G. and Gal, T. (2001). Redistribution of funds for teaching and research among universities: the case of North Rhine-Westphalia, *European Journal of Operational Research* **130**, 1, pp. 111–120.

Feng, J. W. and Yu, P. L. (1998). Minimum spanning table and optimal expansion of competence set, *Journal of Optimization Theory and Applications* **99**, 3, pp. 655–679.

Fikes, R., Farquhar, A. and Rice, J. (1997). Tools for assembling modular ontologies in Ontolingua, in *Proceedings of AAAI 97*, pp. 436–441.

Filipowicz, B. (1996). *Modele stochastyczne w badaniach operacyjnych: analiza i synteza systemów obsługi i sieci kolejkowych* (Wydawnictwa Naukowo-Techniczne).

Freeman, L. A. (2001). Information systems knowledge: foundations, definitions, and applications, *Information Systems Frontiers* **3**, 2, pp. 249–266.

Friesen, N. and McGreal, R. (2002). International e-learning specifications, *The International Review of Research in Open and Distributed Learning* **3**, 2.

Galwas, B. (2003). Technika prowadzenia przedmiotu przez Internet, in *Materiały z III Konferencji i Warsztatów Uniwersytet Wirtualny: Model, Narzędzia i Praktyka, Warszawa*.

Gobet, F. and Clarkson, G. (2004). Chunks in expert memory: evidence for the magical number four... or is it two? *Memory* **12**, 6, pp. 732–747.

Gobet, F., Lane, P. C. R., Croker, S., Cheng, P. C. H., Jones, G., Oliver, I. and Pine, J. M. (2001). Chunking mechanisms in human learning, *Trends in Cognitive Sciences* **5**, 6, pp. 236–243.

Gobet, F. and Simon, H. A. (2000). Five seconds or sixty? Presentation time in expert memory, *Cognitive Science* **24**, 4, pp. 651–682.

Goldstein, R. C. and Storey, V. C. (1999). Data abstractions: why and how? *Data & Knowledge Engineering* **29**, 3, pp. 293–311.

Gómez, A., Moreno, A., Pazos, J. and Sierra-Alonso, A. (2000). Knowledge maps: an essential technique for conceptualisation, *Data & Knowledge Engineering* **33**, 2, pp. 169–190.

Gómez, T., Gonzalez, M., Luque, M., Miguel, F. and Ruiz, F. (2001). Multiple objectives decomposition–coordination methods for hierarchical organizations, *European Journal of Operational Research* **133**, 2, pp. 323–341.

Gómez-Pérez, A., Fernández-López, M. and Corcho, O. (2004). *Ontological Engineering: With Examples from the Areas of Knowledge Management, e-Commerce and the Semantic Web* (Springer).

Gordon, J. L. (2000). Creating knowledge maps by exploiting dependent relationships, in *Applications and Innovations in Intelligent Systems VII* (Springer), pp. 64–78.

Gordon, W. J. and Newell, G. F. (1967). Closed queuing systems with exponential servers, *Operations Research* **15**, 2, pp. 254–265.

Graves, S. C., Kletter, D. B. and Hetzel, W. B. (1998). A dynamic model for requirements planning with application to supply chain optimization, *Operations Research* **46**, 3-supplement-3, pp. S35–S49.

Greenberg, L. (2002). LMS and LCMS: what's the difference, *Learning Circuits* **31**, 2.

Gruber, T. (2009). *Ontology* (Springer, Boston, MA), ISBN 978-0-387-39940-9, pp. 1963–1965, doi:10.1007/978-0-387-39940-9_1318, https://doi.org/10.1007/978-0-387-39940-9_1318

Guarino, N. (1997). Understanding, building and using ontologies, *International Journal of Human-Computer Studies* **46**, 2-3, pp. 293–310.

Guarino, N. (1998). Formal ontology in information systems, in *Proceedings of the First International Conference (FOIS'98), June 6–8, Trento, Italy* (IOS press, Amsterdam), pp. 3–15.

Gubko, M. V. and Novikov, D. A. (2002). *Game Theory in Control of Organizational Systems (in Russian)* (Sinteg).

Hamel, C. J. and Ryan-Jones, D. (2002). Designing instruction with learning objects, *International Journal of Educational Technology* **3**, 1, pp. 111–124.

Hamel, C. J. and Ryan-Jones, D. L. (2001). We're Not Designing Courses Anymore, World Conference on the WWW and Internet Proceedings. Orlando, FL, October 23–27, 2001.

Hartnett, M., George, A. S. and Dron, J. (2011). Examining motivation in online distance learning environments: complex, multifaceted and situation-dependent, *International Review of Research in Open and Distributed Learning* **12**, 6, pp. 20–38.

Helic, D., Maurer, H. and Scerbakov, N. (2004). Knowledge transfer processes in a modern WBT system, *Journal of Network and Computer Applications* **27**, 3, pp. 163–190.

Henderson-Sellers, B., Collins, G., Dué, R. and Graham, I. (2001). A qualitative comparison of two processes for object-oriented software development, *Information and Software Technology* **43**, 12, pp. 705–724.

Hestenes, D. (1995). Modeling software for learning and doing physics, in *Thinking Physics for Teaching* (Springer), pp. 25–65.

Heylighen, F. (1990). *Representation and Change: A Metarepresentational Framework for the Foundations of Physical and Cognitive Science* (Communication & Cognition).

Heywood, L., Gonczi, A. and Hager, P. (1992). *A Guide to Development of Competency Standards for Professions* (Australian Government Publishing Service, Canberra).

Hilburn, T. B., Hislop, G., Bagert, D. J., Lutz, M., Mengel, S. and McCracken, M. (1999). Guidance for the development of software engineering education programs, *Journal of Systems and Software* **49**, 2-3, pp. 163–169.

Hitt, M. A., Ireland, R. D. and Hoskisson, R. E. (2005). *Strategic Management: Competitiveness and Globalization: Cases* (South-Western College Pub, Cincinnati, Ohio), ISBN 0324275331, includes bibliographical references.

Hofrichter, D. A. and Spencer Jr, L. M. (1996). Competencies: the right foundation the right foundation for effective human resources management, *Compensation & Benefits Review* **28**, 6, pp. 21–26.

HR-XML Consortium (2006). Competencies (measurable characteristics), recommendation, 2006 feb 28, http://xml.coverpages.org/HR-XML-Competencies-1_0.pdf

Hrastinski, S. (2007). *Participating in Synchronous Online Education*, Vol. 6 (Lund University).

Hutmacher, W. (1997). Key Competencies for Europe. Report of the Symposium (Berne, Switzerland, March 27-30, 1996). A Secondary Education for Europe Project. Council for Cultural Cooperation, Strasbourg (France).

IEEE LTSC (2006). Proposed Draft Standard for Learning Technology: Simple Reusable Competency Map, Rev. 4, https://ieeeltsc.files.wordpress.com/2009/03/reusablecompetencymapproposal.pdf

IEEE LTSC (2007). P1484.20.1 Reusable Competency Definitions Draft 8, https://www.ieeeltsc.org/working-groups/wg20Comp/wg20rcdfolder/IEEE_1484.20.1.D8.pdf/view/

ISO/IEC (2000). 2000 Information Technology: SGML Applications; Topic Maps (ISO/IEC 13250:2000), International Organization for Standardization, Geneva, Switzerland.

Johnes, J. (1996). Performance assessment in higher education in Britain, *European Journal of Operational Research* **89**, 1, pp. 18–33.

Juszczyk, S. (2002). *Edukacja na odległość: kodyfikacja reguł i procesów* (Adam Marszałek).

Kałuski, J. (2002). *Teoria Gier* (Wydawawnictwo Politechniki Śląskiej).

Kassanke, S., El-Saddik, A. and Steinacker, A. (2001). Learning objects metadata and tools in the area of operations research, in *Proceedings of InED-Media 2001 World Conference on Educational Multimedia, Hypermedia & Telecommunications* (Tampere, Finland), pp. 891–895.

Keller, J. M. (1999). Using the arcs motivational process in computer-based instruction and distance education, *New Directions for Teaching and Learning* **1999**, 78, pp. 37–47.

Kelting-Gibson, L. M. (2005). Comparison of curriculum development Practices, *Educational Research Quarterly* **29**, 1, pp. 26–36.

Kelton, W. D., Sadowski, R. P. and Sadowski, D. A. (1997). *Simulation with Arena* (McGraw-Hill).

Kent, R. E. (2000). Conceptual knowledge markup language: an introduction, *Netnomics* **2**, 2, pp. 139–169.

Kent, R. E. (2011). A KIF formalization for the IFF category theory ontology, *arXiv preprint arXiv:1109.0333* .

Khalifa, M. and Lam, R. (2002). Web-based learning: effects on learning process and outcome, *IEEE Transactions on Education* **45**, 4, pp. 350–356.

Kleinrock, L. (1975). *Queueing Systems. Volume I: Theory* (Wiley).

Knudsen, C. J. S. and Naeve, A. (2002). Presence production in a distributed shared virtual environment for exploring mathematics, in *Advanced Computer Systems* (Springer), pp. 149–159.

Kofoed, L. B. and Stachowicz Marian, S. (2012). Problem-based learning principles in two pedagogical models, in *Proceedings from ICIT Conference on Information and Communication Technologies in Education, Manufacturing and Research.*

Korytkowski, P. and Zaikin, O. (2004). Zarządzanie zdolnością produkcyjną w produkcji niematerialnej, in R. e. a. Kulikowaki (ed.), *Badania Operacyjne i Systemowe* (Exit, Warszawa), pp. 207–218.

Kushtina, E. (2006). Koncepcja otwartego systemu informacyjnego nauczania zdalnego, *Prace Naukowe Politechniki Szczecińskiej. Wydział Informatyki*, 589 (1), pp. 1–163.

Kushtina, E., Dolgui, A. and Malachowski, B. (2005). Organization of the modeling and simulation of the discrete processes, in *Information Processing and Security Systems* (Springer), pp. 443–452.

Kushtina, E. and Różewski, P. (2003). Opracowanie podejscia do tworzenia formalnego opisu dziedzinowej wiedzy teoretycznej, *Badania Systemowe, IBS PAN* **33**, pp. 29–40.

Kushtina, E. and Różewski, P. (2004). Analiza systemowa idei otwartego nauczania zdalnego, in A. Straszak and J. Owsiński (eds.), *Badania Operacyjne i Systemowe* (EXIT), pp. 231–245.

Kushtina, E., Zaikin, O. and Enlund, N. (2001). A knowledge base approach to courseware design for distance learning, in *EUNIS'01*, pp. 499–505.

Kushtina, E., Zaikin, O. and Rózewski, P. (2007). On the knowledge repository design and management in e-learning, in J. Lu, G. Zhang and D. Ruan (eds.), *E-Service Intelligence: Methodologies, Technologies and Applications* (Springer), pp. 497–515.

Kushtina, E., Zaikin, O., Różewski, P. and Małachowski, B. (2009). Cost estimation algorithm and decision-making model for curriculum modification in educational organization, *European Journal of Operational Research* **197**, 2, pp. 752–763.

Kusztina, E., Zaikin, O. and Tadeusiewicz, R. (2010). The research behavior/attitude support model in open learning systems, *Bulletin of the Polish Academy of Sciences: Technical Sciences* **58**, 4, pp. 705–711.

Kwak, N. K. and Lee, C. (1998). A multicriteria decision-making approach to university resource allocations and information infrastructure planning, *European Journal of Operational Research* **110**, 2, pp. 234–242.

Larreamendy-Joerns, J. and Lainhardt, G. (2006). Going the distance with online education. *Review of Educational Research* **76**(4), 567–605, doi: 10.3102/00346543076004567.

Law, A. M. and Kelton, W. D. (2000). *Simulation Modeling and Analysis* (McGraw-Hill).

Li, H.-L. (1999). Incorporating competence sets of decision makers by deduction graphs, *Operations Research* **47**, 2, pp. 209–220.

Li, J.-M., Chiang, C.-I. and Yu, P.-L. (2000). Optimal multiple stage expansion of competence set, *European Journal of Operational Research* **120**, 3, pp. 511–524.

Lin, C.-C. (2006). Competence set expansion using an efficient 0-1 programming model, *European Journal of Operational Research* **170**, 3, pp. 950–956.

Lin, Y. T., Tseng, S.-S. and Tsai, C.-F. (2003). Design and implementation of new object-oriented rule base management system, *Expert Systems with Applications* **25**, 3, pp. 369–385.

Liu, Y. and Ginther, D. (1999). Cognitive styles and distance education, *Online Journal of Distance Learning Administration* **2**, 3.

Malawski, M., Wieczorek, A. and Sosnowska, H. (2004). Teoria Gier, *PWN, Warszawa*.

Malinowska, M., Kusztina, E., Zaikin, O., Reng, L., Kofoed, L. B. and Żyławski, A. (2013). Modeling the competence acquiring process in higher education institution, *IFAC Proceedings Volumes* **46**, 9, pp. 1578–1583.

Mansfield, R. S. (1996). Building competency models: approaches for HR professionals, *Human Resource Management* **35**, 1, pp. 7–18.

Maruszewski, T. (2002). *Psychologia Poznania [Psychology of Cognition]* (Gdańskie Wydawnictwo Psychologiczne, Gdańsk).

McGorry, S. Y. (2003). Measuring quality in online programs, *The Internet and Higher Education* **6**, 2, pp. 159–177, doi:https://doi.org/10.1016/S1096-7516(03)00022-8, http://www.sciencedirect.com/science/article/pii/S1096751603000228

Mesarovic, M. D., Macko, D. and Takahara, Y. (1970). Theory of hierarchical, multilevel systems, *Mathematics in Science and Engineering* (Academic Press, New York, NY), http://cds.cern.ch/record/233503

Miklashevich, I. A. and Barkaline, V. V. (2005). Mathematical representations of the dynamics of social system: I. General description, *Chaos, Solitons & Fractals* **23**, 1, pp. 195–206.

Miller, R. B. and Brickman, S. J. (2004). A model of future-oriented motivation and self-regulation, *Educational Psychology Review* **16**, 1, pp. 9–33.

Mineau, G. W., Missaoui, R. and Godinx, R. (2000). Conceptual modeling for data and knowledge management, *Data & Knowledge Engineering* **33**, 2, pp. 137–168.

MNiSW (2007). Standardy Kształcenia, http://www.rgnisw.nauka.gov.pl/inne-dokumenty-2007/standardy-ksztalcenia.html

MRPiPS (2016). Krajowe Standardy Kwalifikacji Zawodowych, ftp://kwalifikacje. praca.gov.pl/

MSC2010 (2010). Mathematics Subject Classification 2010, http://msc2010.org/mediawiki/index.php?title=MSC2010

Mulawka, J. J. (1997). *Systemy Ekspertowe* (Wydawnictwa Naukowo-Techniczne).

Nash, J. F. (1950). Equilibrium points in n-person games, *Proceedings of the National Academy of Sciences* **36**, 1, pp. 48–49.

Neches, R., Fikes, R. E., Finin, T., Gruber, T., Patil, R., Senator, T. and Swartout, W. R. (1991). Enabling technology for knowledge sharing, *AI Magazine* **12**, 3, pp. 36–36.

Ng, C.-H. (1996). *Queueing Modelling Fundamentals* (Wiley).

Novikov, D. (2012). *Theory of Control in Organizations* (Nova).

Owen, G. (1975). *Teoria Gier (Game Theory)* (Państwowe Wydawawnictwo Naukowe).

Partington, D., Pellegrinelli, S. and Young, M. (2005). Attributes and levels of programme management competence: an interpretive study, *International Journal of Project Management* **23**, 2, pp. 87–95.

Patru, M. and Khvilon, E. (2002). Open and distance learning: trends, policy and strategy considerations, UNESDOC Digital Library. Programme and meeting document.

Perrenoud, P. (1997). *Construire des compétences dès l'école*, Pratiques et enjeux pédagogiques (ESF éd.), ISBN 9782710112501, https://books.google.pl/books?id=7p7OPQAACAAJ

Phalet, K., Andriessen, I. and Lens, W. (2004). How future goals enhance motivation and learning in multicultural classrooms, *Educational Psychology Review* **16**, 1, pp. 59–89.

PMI (2004). *A Guide to the Project Management Body of Knowledge (PMBOK Guide)* (Project Management Institute, USA).

Polsani, P. R. (2003). Use and abuse of reusable learning objects, *Journal of Digital Information* **3**, 4.

Popov, O., Barcz, A., Piela, P. and Sobczak, T. (2003). Practical realization of modelling an airplane for an intelligent tutoring system, in *Artificial Intelligence and Security in Computing Systems* (Springer), pp. 127–135.

Prabhakar, B., Litecky, C. R. and Arnett, K. (2005). IT skills in a tough job market, *Communications of the ACM* **48**, 10, pp. 91–94.

PROLEARN (2004). Network of Excellence in Professional Learning, http://www.prolearn-project.org/

Quillian, M. R. (1968). Semantic memory, in M. Minsky (ed.), *Semantic Information Processing* (MIT Press), pp. 227–270.

Radosinski, E. (2001). *Systemy Informacyjne w Dynamicznej Analizie Decyzyjnej* (Wydawnictwo Naukowe PWN, Warszawa).

Robinson, S. (2004). *Simulation: The Practice of Model Development and Use* (John Wiley & Sons).

Romainville, M. (1996). L'irrésistible ascension du terme "compétences" en éducation, *Enjeux* **37/38**, pp. 132–142.

Różewski, P. (2004). *Metoda Projektowania Systemu Informatycznego Reprezentacji i Przekazywania Wiedzy dla Nauczania Zdalnego*, PhD thesis, Politechnika Szczecińska, Wydział Informatyki, Szczecin.

Różewski, P., Kusztina, E., Tadeusiewicz, R. and Zaikin, O. (2011). *Intelligent Open Learning Systems: Concepts, Models and Algorithms*, Vol. 22 (Springer Science & Business Media).

Różewski, P. and Zaikin, O. (2015). Integrated mathematical model of competence-based learning-teaching process, *Bulletin of the Polish Academy of Sciences Technical Sciences* **63**, 1, pp. 245–259.

Rumble, G. and Latchem, C. (2003). Organisational models for open and distance learning, in H. Lentell and H. Perraton (eds.), *Policy for Open and Distance Learning* (Routledge), pp. 117–140.

Russel, S. and Norvig, P. (2011). *Artificial Intelligence: A Modern Approach* (Pearson Education).

Rychen, D. S. and Salganik, L. H. (2001). *Defining and Selecting Key Competencies* (Hogrefe & Huber), ISBN 9780889372481.

Sanchez, R. (2004). Understanding competence-based management: identifying and managing five modes of competence, *Journal of Business Research* **57**, 5, pp. 518–532.

Santos, O. A. and Ramos, F. M. S. (2004). Proposal of a framework for Internet-based licensing of learning objects, *Computers & Education* **42**, 3, pp. 227–242.

Saunders, M. and Machell, J. (2000). Understanding emerging trends in higher education curricula and work connections, *Higher Education Policy* **13**, 3, pp. 287–302.

Shannon, R. E. (1978). *Systems Simulation: The Art and Science (in Russian)* (Mir).

Shapley, L. S. (1953). A value for *n*-person games, *Contributions to the Theory of Games* **2**, 28, pp. 307–317.

Sharples, M., Jeffery, N., Du Boulay, J. B. H., Teather, D., Teather, B. and Du Boulay, G. H. (2002). Socio-cognitive engineering: a methodology for the design of human-centred technology, *European Journal of Operational Research* **136**, 2, pp. 310–323.

Sheremetov, L. and Arenas, A. G. (2002). EVA: an interactive web-based collaborative learning environment, *Computers & Education* **39**, 2, pp. 161–182.

Shi, D. S. and Yu, P. L. (1999). Optimal expansion of competence sets with intermediate skills and compound nodes, *Journal of Optimization Theory and Applications* **102**, 3, pp. 643–657, doi:10.1023/A:1022654207989, https://doi.org/10.1023/A:1022654207989

Shubik, M. (1991). *Game Theory in the Social Sciences: Concepts and Solutions* (MIT Press).

Siciński, M. (2003). Kwalifikacje czy kompetencje, *Edukacja i Dialog* **9**, p. 152.

Simon, H. A. (1974). How big is a chunk? *Science* **183**, 4124, pp. 482–488.

Skinner, B. F. (1953). *Science and Human Behavior* (Macmillan).

Small, R. V. and Venkatesh, M. (2000). A cognitive-motivational model of decision satisfaction, *Instructional Science* **28**, 1, pp. 1–22.

Smith, B. (1976). *The Ontology of Reference: Studies in Logic and Phenomenology*, PhD thesis, University of Manchester, Manchester.

Smoczyńska, A. (2005). Kompetencje kluczowe, *Realizacja Koncepcji na Poziomie Szkolnictwa Obowiązkowego*.

Solow, D., Piderit, S., Burnetas, A. and Leenawong, C. (2005). Mathematical models for studying the value of motivational leadership in teams, *Computational & Mathematical Organization Theory* **11**, 1, pp. 5–36.

Sovetov, B. Y. and Yakovlev, S. A. (1998). *System Modeling: Textbook for Universities* (Vysshaya Shkola).

Sowa, J. F. and Zachman, J. A. (1992). Extending and formalizing the framework for information systems architecture, *IBM Systems Journal* **31**, 3, pp. 590–616.

Spencer, L. M. and Spencer, S. M. (1993). *Competence at Work: Models for Superior Performance* (Wiley, New York).

Storey, V. C. (1993). Understanding semantic relationships, *The VLDB Journal* **2**, 4, pp. 455–488.

Straffin, P. D. (2004). *Teoria Gier* (Scholar).

Studer, R., Benjamins, V. R. and Fensel, D. (1998). Knowledge engineering: principles and methods, *Data & Knowledge Engineering* **25**, 1-2, pp. 161–197.

Sugumaran, V. and Storey, V. C. (2002). Ontologies for conceptual modeling: their creation, use, and management, *Data & Knowledge Engineering* **42**, 3, pp. 251–271.

Sure, Y., Maedche, A. and Staab, S. (2000). Leveraging corporate skill knowledge: from ProPer to OntoProPer, in *PAKM*, pp. 1–9.

Tait, A. (1996). Open and distance learning policy in the European Union 1985–1995, *Higher Education Policy* **9**, 3, pp. 221–238.

Tariq, V. N., Scott, E. M., Cochrane, A. C., Lee, M. and Ryles, L. (2004). Auditing and mapping key skills within university curricula, *Quality Assurance in Education* **12**, 2, pp. 70–81.

Tavares, L. V. (1995). On the development of educational policies, *European Journal of Operational Research* **82**, 3, pp. 409–421.

Tencompetence (2009). Tencompetence: Building the European Network for Lifelong Competence Development, https://cordis. europa. eu/project/id/027087

Trigueiros, M., Cardoso, E. *et al.* (2006). A methodology for the creation and restructuring higher education degrees conformed to the bologna goals: a practical case study, in *12th International Conference of European University Information Systems (EUNIS 2006)*, pp. 176–185.

Tsichritzis, D. C. and Lochovsky, F. H. (1982). *Data Models* (Prentice Hall Professional Technical Reference).

Tuckman, B. W. (2007). The effect of motivational scaffolding on procrastinators' distance learning outcomes, *Computers & Education* **49**, 2, pp. 414–422.

Tuning (2007). Educational Structures in Europe Project (2001–2004), http://unideusto.org/tuningeu/

Turner, D. and Crawford, M. (1994). Managing current and future competitive performers: the role of competency, in G. Hamel and A. Heene (eds.), *Competency-Based Competition*, Strategic Management Series (Wiley), pp. 241–254.

Visser, P. R. S., Jones, D. M., Bench-Capon, T. J. M. and Shave, M. J. R. (1997). An analysis of ontology mismatches; heterogeneity versus interoperability, in *AAAI 1997 Spring Symposium on Ontological Engineering, Stanford CA., USA*, pp. 164–72.

Vogel, E. K. and Machizawa, M. G. (2004). Neural activity predicts individual differences in visual working memory capacity, *Nature* **428**, 6984, pp. 748–751.

Wang, H.-F. and Wang, C. H. (1998). Modelling of optimal expansion of a fuzzy competence set, *International Transactions in Operational Research* **5**, 5, pp. 413–424.

Woodruffe, C. (1992). What is meant by a competency, in R. Boam and P. Sparrow (eds.), *Designing and Achieving Competency: A Competency-*

Based Approach to Developing People and Organization (McGraw-Hill Book Company).

Woolfolk, A. (2012). *Educational Psychology and Classroom Assessment* (Pearson).

Wu, C.-H. (2004). Building knowledge structures for online instructional/learning systems via knowledge elements interrelations, *Expert Systems with Applications* **26**, 3, pp. 311–319.

Yu, P.-L. (1990). *Forming Winning Strategies: An Integrated Theory of Habitual Domains* (Springer-Verlag).

Yu, P.-L. (1991). Habitual domains, *Operations Research* **39**, 6, pp. 869–876.

Yu, P. L. and Zhang, D. (1990). A foundation for competence set analysis, *Mathematical Social Sciences* **20**, 3, pp. 251–299, doi:https://doi.org/10.1016/0165-4896(90)90005-R, http://www.sciencedirect.com/science/article/pii/016548969090005R

Zadeh, L. A. (1965). Fuzzy sets, *Information and Control* **8**, 3, pp. 338–353.

Zaikin, O. (2002a). *Queuing Modelling of Supply Chain in Intelligent Production* (Faculty of Computer Science and Information Systems, Technical University of Szczecin).

Zaikin, O. (2002b). Resource distribution in automatic production control for nonmaterial products: a mathematical model, *Automation and Remote Control* **63**, 8, pp. 1351–1356.

Zaikin, O. and Dolgui, A. (2000). Simulation model for optimization of resources allocation in the queuing networks, in I. Troch and F. Breitenecker (eds.), *Proceedings 3rd MATHMOD, IMACS: Symposium on Mathematical Modelling* (Wien).

Zaikin, O., Kushtina, E. and Różewski, P. (2006). Model and algorithm of the conceptual scheme formation for knowledge domain in distance learning, *European Journal of Operational Research* **175**, 3, pp. 1379–1399.

Zaikin, O., Kushtina, E., Różewski, P., Maslowski, M., Holma, J., Snellman, M., Junes, S. and Ruottinen, S. (2004). eQuality project: the quality aspect of teaching and learning in student life-cycle, in *Proceedings of the Technology Enhanced Learning Conference (T.E.L'04)* (Milano, Italy).

Zaikin, O., Kusthina, E. and Różewsk, P. (2000). Analiza wykorzystania sieci telekomunikacyjnych w zastosowaniach nauczania na odleglosc, in *Materialy z V Poznanskich Warsztatow Telekomunikacyjnych* (Politechnika Poznanska, Poznan), pp. 1–4.

Zaikin, O., Malinowska, M., Kofoed, L. B., Tadeusiewicz, R. and Żyławski, A. (2014). The behavioural motivation model in open distance learning, in *COGNITIVE 2014, The Sixth International Conference on Advanced Cognitive Technologies and Applications* (IARIA), pp. 226–234.

Zaikin, O. and Różewski, P. (2005). Wirtualne laboratorium symulacji procesow produkcyjnych: program e-quality, in *I Konferencja Nowe Technologie w Ksztalceniu na Odleglosc* (Koszalin - Osieki), pp. 287–296.

Zaikin, O., Tadeusiewicz, R., Różewski, P., Kofoed, L. B., Malinowska, M. and Żyławski, A. (2016). Teachers' and students' motivation model as a strategy for open distance learning processes, *Bulletin of the Polish Academy of Sciences Technical Sciences* **64**, 4, pp. 943–955.

Zaliwski, A. (2000). *Korporacyjne Bazy Wiedzy* (Polskie Wydawnictwo Ekonomiczne).

Zawacki-Richter, O. and Anderson, T. (2014). *Online Distance Education: Towards a Research Agenda* (Athabasca University Press).

Index